Maximize Your Lab Time!

Access included with every new book.

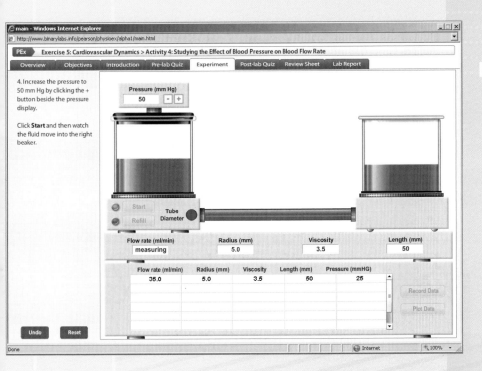

Registration will let you:

- Perform 63 online labs
- Review basic concepts
- Assess your learning with Pre-Lab and Post-Lab Quizzes and Review Sheet Questions
- Save all of your results in PDF Lab Report

www.physioex.com

TO REGISTER

1. Go to www.physioex.com
2. Click the "Register."
3. Follow on-screen instructions to create your login name and password.

Your Access Code is:

USSPEA-CLVII-SATYR-DISCS-FRACK-CHEWA

*

Note: If there is no silver foil covering the access code, it may already have been redeemed, and therefore may no longer be valid. In that case, you can purchase access online using a major credit card or PayPal account. To do so, go to www.physioex.com, click on "Buy Access," and follow the on-screen instructions.

TO LOG IN

1. Go to www.physioex.com
2. Click "Log In."
3. Pick your book cover.
4. Enter your login name and password.
5. Click "Log In."

Hint:
Remember to bookmark the site after you log in.

Technical Support:
http://247pearsoned.custhelp.com

PhysioEx™ 9.1
Laboratory Simulations
in Physiology

Peter Zao
North Idaho College

Timothy Stabler
Indiana University Northwest

Lori Smith
American River College

Andrew Lokuta
University of Wisconsin—Madison

Edwin Griff
University of Cincinnati

PEARSON

Boston Columbus Indianapolis New York San Francisco Upper Saddle River
Amsterdam Cape Town Dubai London Madrid Milan Munich Paris Montréal Toronto
Delhi Mexico City São Paulo Sydney Hong Kong Seoul Singapore Taipei Tokyo

Editor-in-Chief: Serina Beauparlant

Associate Project Editor: Shannon Cutt

Development Manager: Barbara Yien

Editorial Assistant: John Maas

Managing Editor: Deborah Cogan

Production Manager: Michele Mangelli

Production Supervisor: Janet Vail

Interior and Cover Designer: Riezebos Holzbaur Design Group

Copyeditor: Anita Wagner

Compositor: S4Carlisle Publishing Services

Media Producer: Erik Fortier

PhysioEx Developer: BinaryLabs, Inc.

Senior Manufacturing Buyer: Stacey Weinberger

Marketing Manager: Derek Perrigo

Cover photograph credit: Pascal Goetgheluck/Photolibrary

ISBN 10: 0-321-90541-5

ISBN 13: 978-0-321-90541-3

Contents

Preface v

Exercise 1 Cell Transport Mechanisms and Permeability 1

Exercise 2 Skeletal Muscle Physiology 15

Exercise 3 Neurophysiology of Nerve Impulses 33

Exercise 4 Endocrine System Physiology 57

Exercise 5 Cardiovascular Dynamics 73

Exercise 6 Cardiovascular Physiology 91

Exercise 7 Respiratory System Mechanics 103

Exercise 8 Chemical and Physical Processes of Digestion 117

Exercise 9 Renal System Physiology 129

Exercise 10 Acid-Base Balance 147

Exercise 11 Blood Analysis 159

Exercise 12 Serological Testing 175

Index 187

Preface

The PhysioEx CD-ROM is easy-to-use laboratory simulation software and lab exercises that consist of 12 exercises containing a total of 63 physiology lab activities. It can be used to supplement or substitute for wet labs. PhysioEx allows students to repeat labs as often as they like, perform experiments without harming live animals, and conduct experiments that are difficult to perform in a wet lab environment because of time, cost, or safety concerns. The PhysioEx 9.0 CD-ROM comes packaged with every new copy of the lab manual and is available at www.physioex.com.

NEW TO VERSION 9.0

The new online format includes:

• **Easy step-by-step instructions** so that everything students need to do to complete the lab is located in one convenient place. Students gather data, analyze results, and check their understanding all on screen.

• **Stop & Think Questions and Predict Questions** help students think about the connection between the activities and the physiological concepts they demonstrate.

• **Greater data variability in the results** reflects more realistically the results students would encounter in a wet lab experiment.

• **New Pre-lab and Post-lab Quizzes,** and short-answer Review Sheets are offered for every activity.

• **Students can save their Lab Report as a PDF,** which they can print and/or email to their instructor.

• **Seven videos of lab experiments** demonstrate the actual experiments simulated on-screen, making it easy for students to understand and visualize the context of the simulations. Videos demonstrate the following experiments: Skeletal Muscle, Blood Typing, Cardiovascular Physiology, Use of a Water-Filled Spirometer, Nerve Impulses, BMR Measurement, and Cell Transport.

The new lab manual includes:

• **New Pre-lab Quizzes** at the beginning of each exercise motivate your students to prepare for lab by testing basic information they should know prior to completing the exercise. These quizzes are different from those found online.

• **New Activity Questions** that challenge students to apply and reinforce what they've learned.

Instructor Resources include:

• **A Test Bank of assignable pre-lab and post-lab quizzes** for use with TestGen® or any course management system provided for instructors.

• **A fully updated Instructor Guide** containing answers to Pre-lab Quizzes, Review Sheets, and Activity Questions. ISBN: 0-321-75095-0 / 978-0-321-75095-2

> **NOTE:** For PhysioEx 9.0, there is one version only of PhysioEx. We have combined the previous A&P and Physiology versions of PhysioEx into one product.

TOPICS IN THIS EDITION

• Exercise 1: *Cell Transport Mechanisms and Permeability.* Explores how substances cross the cell's membrane. Topics covered include: simple and facilitated diffusion, osmosis, filtration, and active transport.

• Exercise 2: *Skeletal Muscle Physiology.* Provides insights into the complex physiology of skeletal muscle. Topics include: electrical stimulation, isometric contractions, and isotonic contractions.

• Exercise 3: *Neurophysiology of Nerve Impulses.* Investigates stimuli that elicit action potentials, stimuli that inhibit action potentials, and factors affecting the conduction velocity of an action potential.

• Exercise 4: *Endocrine System Physiology.* Investigates the relationship between hormones and metabolism; the effect of estrogen replacement therapy; the diagnosis of diabetes; and the relationship between the levels of cortisol and adrenocorticotropic hormone and a variety of endocrine disorders.

• Exercise 5: *Cardiovascular Dynamics.* Allows students to perform experiments that would be difficult if not impossible to do in a traditional laboratory. Topics include vessel resistance and pump (heart) mechanics.

• Exercise 6: *Cardiovascular Physiology.* Examines variables influencing heart activity. Topics include: setting up and recording baseline heart activity, the refractory period of cardiac muscle, and an investigation of factors that affect heart rate and contractility.

• Exercise 7: *Respiratory System Mechanics.* Investigates physical and chemical aspects of pulmonary function. Students collect data simulating normal lung volumes. Other activities examine factors such as airway resistance and the effect of surfactant on lung function.

• Exercise 8: *Chemical and Physical Processes of Digestion.* Examines factors that affect enzyme activity by manipulating (in compressed time) enzymes, reagents, and incubation conditions.

• Exercise 9: *Renal System Physiology.* Simulates the function of a single nephron. Topics include: factors influencing glomerular filtration, the effect of hormones on urine function, and glucose transport maximum.

• Exercise 10: *Acid-Base Balance.* Topics include: respiratory and metabolic acidosis/alkalosis, and renal and respiratory compensation.

• Exercise 11: *Blood Analysis.* Topics include: hematocrit determination, erythrocyte sedimentation rate determination, hemoglobin determination, blood typing, and total cholesterol determination.

• Exercise 12: *Serological Testing.* Investigates antigen-antibody reactions and their role in clinical tests used to diagnose a disease or an infection.

ACKNOWLEDGMENTS

We wish to thank the following reviewers for their contributions to this edition: Beth Altschafl, University of Wisconsin–Madison; Isaac Barjis, NYC College of Technology; Claudie Biggers, Amarillo College; William Boyko, Sinclair Community College; Carol Bunde, Idaho State University; Steve Burnett, Clayton State University; Sara Colosimo, University of North Florida; Andrew Criss, CUNY York College; Chester Cooper, Odessa College; Lynnette Danzl-Tauer, Rock Valley College; Jake Dechant, University of Pittsburgh; Richard Doolin, Daytona State College; Joseph Esdin, UCLA; Robert Eveslage, Cincinnati State University; Terry Flum, Southern State Community College; Durwood Foote, Tarrant County College; Chris Gan, Highline Community College; Anthony Gaudin, Ivy Tech Community College; Mike Gilbert, Fresno City College; Terri Gillian, Virginia Tech University; Lauren Gollahon, Texas Tech University; Barbara Heard, Atlantic Cape Community College; Rodney Holmes, Waubonsee Community College; Alexander Ibe, Weatherford College; Jody Johnson, Arapahoe Community College; Cindy Jones, Colorado Community College Online; Steve Kash, Oklahoma City Community College; Suzanne Kempke, St. Johns River Community College; Will Kleinelp, Middlesex Community College; Linda Kollett, Massasoit Community College; Bonnie McCormick, University of Incarnate Word; Tammy McNutt-Scott, Clemson University; Susan Mitchell, Onondaga Community College; Renee Moore, Solano Community College; Lynn Preston, Tarrant County College; Julie Richmond, University of North Florida; Jo Rogers, University of Cincinnati; Nidia Romer, Miami Dade College; William Rose, University of Delaware; Mark Schmidt, Clark State Community College; Amira Shaham-Albalancy, Collin County Community College; Denise Slayback-Barry, Indiana University–Purdue University; Phillip Snider, Gadsden State Community College; Cindy Stanfield, University of South Alabama; Dianne Synder, Augusta State University; Laura Steele, Ivy Tech–Fort Wayne; Dieterich Steinmetz, Portland Community College; Lynn Sweeney, San Juan College; Emily Taylor, Cal Polytechnic State University; Charles Venglarik, Jefferson State Community College; Paul Wagner, Washburn Community College; Janice Webster, Ivy Tech Community College; Marlena West, Madisonville Community College; Jane Wiggens, Middlesex Community College; and Scott Zimmerman, Missouri State University.

Special thanks go to Sandra Stewart of Vincennes University for her contributions and commitment to checking the accuracy of the software. Thank you also to Josephine Rogers of the University of Cincinnati for authoring the brand new pre-lab quizzes that have been adapted from the Marieb/Mitchell *Human Anatomy & Physiology Laboratory Manual*, Tenth Edition Update.

The excellence of PhysioEx 9.0 reflects the expertise of Peter Zao, Timothy Stabler, Lori Smith, Andrew Lokuta, Greta Peterson, Nina Zanetti, and Edwin Griff. They generated the ideas behind the activities and simulations. Credit also goes to the team at BinaryLabs, Inc., for their expert programming and design.

Last but not least, thanks to our colleagues and friends at Pearson-Benjamin Cummings who worked on PhysioEx 9.0, especially Serina Beauparlant, Editor-in-Chief; Erik Fortier, Media Producer; and Shannon Cutt, Associate Project Editor. Many thanks to Stacey Weinberger for her manufacturing expertise and to our Marketing Manager, Derek Perrigo. Thank you also to Michele Mangelli and Janet Vail, who expertly managed production of the printed lab manual. Our beautiful interior and cover designs were created by Yvo Riezebos. Anita Wagner was our thorough copyeditor.

To Joseph Thomas Nason (1938–2010); the only person I have ever known who could readily identify words that use "w" as a vowel (e.g. crwth).
—Andrew Lokuta

To Paul, my husband, and Jared, my son; for your support, understanding, and inspiration.
—Lori Smith

Cell Transport Mechanisms and Permeability

Exercise Overview

The molecular composition of the plasma membrane allows it to be selective about what passes through it. It allows nutrients and appropriate amounts of ions to enter the cell and keeps out undesirable substances. For that reason, we say the plasma membrane is **selectively permeable.** Valuable cell proteins and other substances are kept within the cell, and metabolic wastes pass to the exterior.

Transport through the plasma membrane occurs in two basic ways: either passively or actively. In **passive processes,** the transport process is driven by concentration or pressure differences *(gradients)* between the interior and exterior of the cell. In **active processes,** the cell provides energy (ATP) to power the transport.

Two key passive processes of membrane transport are **diffusion** and **filtration.** Diffusion is an important transport process for every cell in the body. **Simple diffusion** occurs without the assistance of membrane proteins, and **facilitated diffusion** requires a membrane-bound carrier protein that assists in the transport.

In both simple and facilitated diffusion, the substance being transported moves *with* (or *along* or *down*) the *concentration gradient* of the solute (from a region of its higher concentration to a region of its lower concentration). The process does not require energy from the cell. Instead, energy in the form of **kinetic energy** comes from the constant motion of the molecules. The movement of solutes continues until the solutes are evenly dispersed throughout the solution. At this point, the solution has reached **equilibrium.**

A special type of diffusion across a membrane is **osmosis.** In osmosis, water moves with its concentration gradient, from a higher concentration of water to a lower concentration of water. It moves in response to a higher concentration of solutes on the other side of a membrane.

In the body, the other key passive process, **filtration,** usually occurs only across capillary walls. Filtration depends upon a *pressure gradient* as its driving force. It is not a selective process. It is dependent upon the size of the pores in the filter.

The two key active processes (recall that active processes require energy) are **active transport** and **vesicular transport.** Like facilitated diffusion, active transport uses a membrane-bound carrier protein. Active transport differs from facilitated diffusion because the solutes move *against* their concentration gradient and because ATP is used to power the transport. Vesicular transport includes phagocytosis, endocytosis, pinocytosis, and exocytosis. These processes are not covered in this exercise. The activities in this exercise will explore the cell transport mechanisms individually.

Simulating Dialysis (Simple Diffusion)

OBJECTIVES

1. To understand that diffusion is a passive process dependent upon a solute concentration gradient.
2. To understand the relationship between molecular weight and molecular size.
3. To understand how solute concentration affects the rate of diffusion.
4. To understand how molecular weight affects the rate of diffusion.

Introduction

Recall that all molecules possess *kinetic energy* and are in constant motion. As molecules move about randomly at high speeds, they collide and bounce off one another, changing direction with each collision. For a given temperature, all matter has about the same average kinetic energy. Smaller molecules tend to move faster than larger molecules because kinetic energy is directly related to both mass and velocity ($KE = \frac{1}{2}mv^2$).

When a **concentration gradient** (difference in concentration) exists, the net effect of this random molecular movement is that the molecules eventually become evenly distributed throughout the environment—in other words, diffusion occurs. **Diffusion** is the movement of molecules from a region of their higher concentration to a region of their lower concentration. The driving force behind diffusion is the kinetic energy of the molecules themselves.

The diffusion of particles into and out of cells is modified by the plasma membrane, which is a physical barrier. In general, molecules diffuse passively through the plasma membrane if they are small enough to pass through its pores (and are aided by an electrical and/or concentration gradient) or if they can dissolve in the lipid portion of the membrane (as in the case of CO_2 and O_2). A membrane is called *selectively permeable, differentially permeable,* or *semipermeable* if it allows some solute particles (molecules) to pass but not others.

The diffusion of *solute particles* dissolved in water through a selectively permeable membrane is called **simple diffusion.** The diffusion of *water* through a differentially permeable membrane is called **osmosis.** Both simple diffusion and osmosis involve movement of a substance from an area of its higher concentration to an area of its lower concentration, that is, *with* (or *along* or *down)* its concentration gradient.

This activity provides information on the passage of water and solutes through selectively permeable membranes. You can apply what you learn to the study of transport mechanisms in living, membrane-bounded cells. The dialysis membranes used each have a different *molecular weight cutoff (MWCO),* indicated by the number below it. You can think of MWCO in terms of pore size: the larger the MWCO number, the larger the pores in the membrane. The molecular weight of a solute is the number of grams per mole, where a mole is the constant Avogadro's number 6.02×10^{23} molecules/mole. The larger the molecular weight, the larger the mass of the molecule. The term molecular mass is sometimes used instead of molecular weight.

> **EQUIPMENT USED** The following equipment will be depicted on-screen: left and right beakers—used for diffusion of solutes; dialysis membranes with various molecular weight cutoffs (MWCOs).

Experiment Instructions

Go to the home page in the PhysioEx software and click **Exercise 1: Cell Transport Mechanisms and Permeability.** Click **Activity 1: Simulating Dialysis (Simple Diffusion),** and take the online **Pre-lab Quiz** for Activity 1.

After you take the online Pre-lab Quiz, click the **Experiment** tab and begin the experiment. The experiment instructions are reprinted here for your reference. The opening screen for the experiment is shown below.

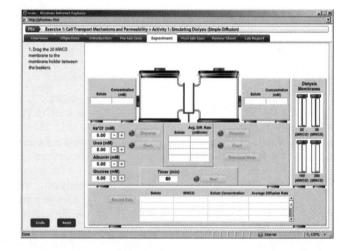

1. Drag the 20 MWCO membrane to the membrane holder between the beakers.

2. Increase the Na^+Cl^- concentration to be dispensed to the left beaker to 9.00 mM by clicking the + button beside the Na^+Cl^- display. Click **Dispense** to fill the left beaker with 9.00 mM Na^+Cl^- solution.

3. Note that the concentration of Na^+Cl^- in the left beaker is displayed in the concentration window to the left of the beaker. Click **Deionized Water** and then click **Dispense** to fill the right beaker with deionized water.

4. After you start the run, the barrier between the beakers will descend, allowing the solutions in each beaker to have access to the dialysis membrane separating them. You will be

able to determine the amount of solute that passes through the membrane by observing the concentration display to the side of each beaker. A level above zero in Na^+Cl^- concentration in the right beaker indicates that Na^+ and Cl^- ions are diffusing from the left beaker into the right beaker through the selectively permeable dialysis membrane. Note that the timer is set to 60 minutes. The simulation compresses the 60-minute time period into 10 seconds of real time. Click **Start** to start the run and watch the concentration display to the side of each beaker for any activity.

5. Click **Record Data** to display your results in the grid (and record your results in Chart 1).

CHART 1	Dialysis Results (average diffusion rate in mM/sec)			
	Membrane MWCO			
Solute	20	50	100	200
Na^+Cl^-				
Urea				
Albumin				
Glucose				

? **PREDICT Question 1**
The molecular weight of urea is 60.07. Do you think urea will diffuse through the 20 MWCO membrane?

_____

6. Click **Flush** beneath each of the beakers to prepare for the next run.

7. Increase the urea concentration to be dispensed to the left beaker to 9.00 mM by clicking the + button beside the urea display. Click **Dispense** to fill the left beaker with 9.00 mM urea solution.

8. Click **Deionized Water** and then click **Dispense** to fill the right beaker with deionized water.

9. Click **Start** to start the run and watch the concentration display to the side of each beaker for any activity.

10. Click **Record Data** to display your results in the grid (and record your results in Chart 1).

11. Click the 20 MWCO membrane in the membrane holder to automatically return it to the membrane cabinet and then click **Flush** beneath each beaker to prepare for the next run.

12. Drag the 50 MWCO membrane to the membrane holder between the beakers. Increase the Na^+Cl^- concentration to be dispensed to the left beaker to 9.00 mM. Click **Dispense** to fill the left beaker with 9.00 mM Na^+Cl^- solution.

13. Click **Deionized Water** and then click **Dispense** to fill the right beaker with deionized water.

14. Click **Start** to start the run and watch the concentration display to the side of each beaker for any activity.

15. Click **Record Data** to display your results in the grid (and record your results in Chart 1).

16. Click **Flush** beneath each of the beakers to prepare for the next run.

17. Increase the Na^+Cl^- concentration to be dispensed to the left beaker to 18.00 mM. Click **Dispense** to fill the left beaker with 18.00 mM Na^+Cl^- solution.

18. Click **Deionized Water** and then click **Dispense** to fill the right beaker with deionized water.

19. Click **Start** to start the run and watch the concentration display to the side of each beaker for any activity.

20. Click **Record Data** to display your results in the grid (and record your results in Chart 1).

21. Click the 50 MWCO membrane in the membrane holder to automatically return it to the membrane cabinet and then click **Flush** beneath each beaker to prepare for the next run.

22. Drag the 100 MWCO membrane to the membrane holder between the beakers. Increase the Na^+Cl^- concentration to be dispensed to the left beaker to 9.00 mM. Click **Dispense** to fill the left beaker with 9.00 mM Na^+Cl^- solution.

23. Click **Deionized Water** and then click **Dispense** to fill the right beaker with deionized water.

24. Click **Start** to start the run and watch the concentration display to the side of each beaker for any activity.

25. Click **Record Data** to display your results in the grid (and record your results in Chart 1).

26. Click **Flush** beneath each of the beakers to prepare for the next run.

27. Increase the urea concentration to be dispensed to the left beaker to 9.00 mM. Click **Dispense** to fill the left beaker with 9.00 mM urea solution.

28. Click **Deionized Water** and then click **Dispense** to fill the right beaker with deionized water.

29. Click **Start** to start the run and watch the concentration display to the side of each beaker for any activity.

30. Click **Record Data** to display your results in the grid (and record your results in Chart 1).

31. Click the 100 MWCO membrane in the membrane holder to automatically return it to the membrane cabinet and then click **Flush** beneath each beaker to prepare for the next run.

? **PREDICT Question 2**
Recall that glucose is a monosaccharide, albumin is a protein with 607 amino acids, and the average molecular weight of a single amino acid is 135 g/mole. Will glucose or albumin be able to diffuse through the 200 MWCO membrane?

_____

32. Drag the 200 MWCO membrane to the membrane holder between the beakers. Increase the glucose concentration to be dispensed to the left beaker to 9.00 m*M*. Click **Dispense** to fill the left beaker with 9.00 m*M* glucose solution.

33. Click **Deionized Water** and then click **Dispense** to fill the right beaker with deionized water.

34. Click **Start** to start the run and watch the concentration display to the side of each beaker for any activity.

35. Click **Record Data** to display your results in the grid (and record your results in Chart 1).

36. Click **Flush** beneath each of the beakers to prepare for the next run.

37. Increase the albumin concentration to be dispensed to the left beaker to 9.00 m*M*. Click **Dispense** to fill the left beaker with 9.00 m*M* albumin solution.

38. Click **Deionized Water** and then click **Dispense** to fill the right beaker with deionized water.

39. Click **Start** to start the run and watch the concentration display to the side of each beaker for any activity.

40. Click **Record Data** to display your results in the grid (and record your results in Chart 1).

After you complete the experiment, take the online **Post-lab Quiz** for Activity 1.

Activity Questions

1. Did any solutes move through the 20 MWCO membrane? Why or why not?

No the molecular weight
is higher than 20

2. Did Na⁺Cl⁻ move through the 50 MWCO membrane?

Yes, it did reach equilibrium
within 10 min.

3. Describe how the size of a molecule (molecular weight) affects its rate of diffusion.

The greater the size then
the slower rate of diffusion.

4. What happened to the rate of diffusion when you increased the Na⁺Cl⁻ solute concentration?

The rate of diffusion
increased from .0150 to .0273

Simulated Facilitated Diffusion

OBJECTIVES

1. To understand that some solutes require a carrier protein to pass through a membrane because of size or solubility limitations.
2. To observe how the concentration of solutes affects the rate of facilitated diffusion.
3. To observe how the number of transport proteins affects the rate of facilitated diffusion.
4. To understand how transport proteins can become saturated.

Introduction

Some molecules are lipid insoluble or too large to pass through pores in the cell's plasma membrane. Instead, they pass through the membrane by a passive transport process called **facilitated diffusion.** For example, sugars, amino acids, and ions are transported by facilitated diffusion. In this form of transport, solutes combine with carrier-protein molecules in the membrane and are then transported *with* (or *along* or *down*) their concentration gradient. The carrier-protein molecules in the membrane might have to change shape slightly to accommodate the solute, but the cell does not have to expend the energy of ATP.

Because facilitated diffusion relies on carrier proteins, solute transport varies with the number of available carrier-protein molecules in the membrane. The carrier proteins can become saturated if too much solute is present and the maximum transport rate is reached. The carrier proteins are embedded in the plasma membrane and act like a shield, protecting the hydrophilic solute from the lipid portions of the membrane.

Facilitated diffusion typically occurs in one direction for a given solute. The greater the concentration difference between one side of the membrane and the other, the greater the rate of facilitated diffusion.

> **EQUIPMENT USED** The following equipment will be depicted on-screen: left and right beakers—used for diffusion of solutes; dialysis membranes with various molecular weight cutoffs (MWCOs); membrane builder—used to build membranes with different numbers of glucose protein carriers.

Experiment Instructions

Go to the home page in the PhysioEx software and click **Exercise 2: Cell Transport Mechanisms and Permeability.** Click **Activity 2: Simulated Facilitated Diffusion,** and take the online **Pre-lab Quiz** for Activity 2.

After you take the online Pre-lab Quiz, click the **Experiment** tab and begin the experiment. The experiment instructions are reprinted here for your reference. The opening screen for the experiment is shown on the following page.

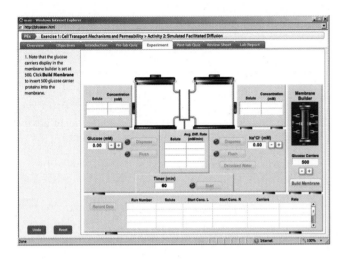

1. Note that the glucose carriers display in the membrane builder is set at 500. Click **Build Membrane** to insert 500 glucose carrier proteins into the membrane.

2. Drag the membrane to the membrane holder between the beakers.

3. Increase the glucose concentration to be dispensed to the left beaker to 2.00 m*M* by clicking the + button beside the glucose display. Click **Dispense** to fill the left beaker with 2.00 m*M* glucose solution.

4. Note that the concentration of glucose in the left beaker is displayed in the concentration window to the left of the beaker. Click **Deionized Water** and then click **Dispense** to fill the right beaker with deionized water.

5. After you start the run, the barrier between the beakers will descend, allowing the solutions in each beaker to have access to the dialysis membrane separating them. You will be able to determine the amount of solute that passes through the membrane by observing the concentration display to the side of each beaker. A level above zero in glucose concentration in the right beaker indicates that glucose is diffusing from the left beaker into the right beaker through the selectively permeable dialysis membrane. Note that the timer is set to 60 minutes. The simulation compresses the 60-minute time period into 10 seconds of real time. Click **Start** to start the run and watch the concentration display to the side of each beaker for any activity.

6. Click **Record Data** to display your results in the grid (and record your results in Chart 2).

CHART 2	Facilitated Diffusion Results (glucose transport rate, mM/sec)		
	Number of glucose carrier proteins		
Glucose concentration	500	700	100
2 m*M*			
8 m*M*			
10 m*M*			
2 m*M* w/ 2.00 m*M* Na+Cl−			

7. Click **Flush** beneath each of the beakers to prepare for the next run.

8. Increase the glucose concentration to be dispensed to the left beaker to 8.00 m*M* by clicking the + button beside the glucose display. Click **Dispense** to fill the left beaker with 8.00 m*M* glucose solution.

9. Click **Deionized Water** and then click **Dispense** to fill the right beaker with deionized water.

10. Click **Start** to start the run and watch the concentration display to the side of each beaker for any activity.

11. Click **Record Data** to display your results in the grid (and record your results in Chart 2).

12. Click the membrane in the membrane holder to automatically return it to the membrane builder and then click **Flush** beneath each beaker to prepare for the next run.

? PREDICT Question 1
What effect do you think increasing the number of protein carriers will have on the glucose transport rate?

13. Increase the number of glucose carriers to 700 by clicking the + button beneath the glucose carriers display. Click **Build Membrane** to insert 700 glucose carrier proteins into the membrane.

14. Drag the membrane to the membrane holder between the beakers. Increase the glucose concentration to be dispensed to the left beaker to 2.00 m*M*. Click **Dispense** to fill the left beaker with 2.00 m*M* glucose solution.

15. Click **Deionized Water** and then click **Dispense** to fill the right beaker with deionized water.

16. Click **Start** to start the run and watch the concentration display to the side of each beaker for any activity.

17. Click **Record Data** to display your results in the grid (and record your results in Chart 2).

18. Click **Flush** beneath each of the beakers to prepare for the next run.

19. Increase the glucose concentration to be dispensed to the left beaker to 8.00 m*M*. Click **Dispense** to fill the left beaker with 8.00 m*M* glucose solution.

20. Click **Deionized Water** and then click **Dispense** to fill the right beaker with deionized water.

21. Click **Start** to start the run and watch the concentration display to the side of each beaker for any activity.

22. Click **Record Data** to display your results in the grid (and record your results in Chart 2).

23. Click the membrane in the membrane holder to automatically return it to the membrane builder and then click **Flush** beneath each beaker to prepare for the next run.

urea 100 MWZJ = .0094 (16 min)

24. Decrease the number of glucose carriers to 100 by clicking the − button beneath the glucose carriers display. Click **Build Membrane** to insert 100 glucose carrier proteins into the membrane.

25. Drag the membrane to the membrane holder between the beakers. Increase the glucose concentration to be dispensed to the left beaker to 10.00 m*M*. Click **Dispense** to fill the left beaker with 10.00 m*M* glucose solution.

26. Click **Deionized Water** and then click **Dispense** to fill the right beaker with deionized water.

27. Click **Start** to start the run and watch the concentration display to the side of each beaker for any activity.

28. Click **Record Data** to display your results in the grid (and record your results in Chart 2).

29. Click the membrane in the membrane holder to automatically return it to the membrane builder and then click **Flush** beneath each beaker to prepare for the next run.

30. Increase the number of glucose carriers to 700. Click **Build Membrane** to insert 700 glucose carrier proteins into the membrane.

> **?** PREDICT **Question 2**
> What effect do you think adding Na⁺Cl⁻ will have on the glucose transport rate?
> _____

31. Increase the glucose concentration to be dispensed to the left beaker to 2.00 m*M*. Click **Dispense** to fill the left beaker with 2.00 m*M* glucose solution.

32. Increase the Na⁺Cl⁻ concentration to be dispensed to the right beaker to 2.00 m*M*. Click **Dispense** button to fill the right beaker with 2.00 m*M* Na⁺Cl⁻ solution.

33. Click **Start** to start the run and watch the concentration display to the side of each beaker for any activity.

34. Click **Record Data** to display the results in the grid (and record your results in Chart 2).

After you complete the experiment, take the online **Post-lab Quiz** for Activity 2.

Activity Questions

1. Are the solutes moving with or against their concentration gradient in facilitated diffusion?

_____

_____

2. What happened to the rate of facilitated diffusion when the number of carrier proteins was increased?

_____

_____

3. Explain why equilibrium was not reached with 10 m*M* glucose and 100 membrane carriers.

_____

_____

_____

4. In the simulation you added Na⁺Cl⁻ to test its effect on glucose diffusion. Explain why there was no effect.

_____

_____

_____ ▄▄

ACTIVITY 3

Simulating Osmotic Pressure

OBJECTIVES

1. To explain how osmosis is a special type of diffusion.
2. To understand that osmosis is a passive process that depends upon the concentration gradient of water.
3. To explain how tonicity of a solution relates to changes in cell volume.
4. To understand conditions that affect osmotic pressure.

Introduction

A special form of diffusion, called **osmosis,** is the diffusion of water through a selectively permeable membrane. (A membrane is called *selectively permeable, differentially permeable,* or *semipermeable* if it allows some molecules to pass but not others.) Because water can pass through the pores of most membranes, it can move from one side of a membrane to the other relatively freely. Osmosis takes place whenever there is a difference in water concentration between the two sides of a membrane.

If we place distilled water on both sides of a membrane, *net* movement of water does not occur. Remember, however, that water molecules would still move between the two sides of the membrane. In such a situation, we would say that there is no *net* osmosis.

The concentration of water in a solution depends on the number of solute particles present. For this reason, increasing the solute concentration coincides with decreasing the water concentration. Because water moves down its concentration gradient (from an area of its higher concentration to an area of its lower concentration), it always moves *toward* the solution with the highest concentration of solutes. Similarly, solutes also move down their concentration gradients.

If we position a *fully* permeable membrane (permeable to solutes and water) between two solutions of differing concentrations, then all substances—solutes and water—diffuse freely, and an equilibrium will be reached between the two sides of the membrane. However, if we use a selectively permeable membrane that is impermeable to the solutes, then we have established a condition where water moves but solutes do not. Consequently, water moves toward the more concentrated solution, resulting in a *volume increase* on that side of the membrane.

By applying this concept to a closed system where volumes cannot change, we can predict that the *pressure* in the more concentrated solution will rise. The force that would need to be applied to oppose the osmosis in a closed system is the **osmotic pressure.** Osmotic pressure is measured in *millimeters of mercury (mm Hg).* In general, the more impermeable the solutes, the higher the osmotic pressure.

Osmotic changes can affect the volume of a cell when it is placed in various solutions. The concept of **tonicity** refers to the way a solution affects the volume of a cell. The tonicity of a solution tells us whether or not a cell will shrink or swell. If the concentration of impermeable solutes is the *same* inside and outside of the cell, the solution is **isotonic.** If there is a *higher* concentration of impermeable solutes *outside* the cell than in the cell's interior, the solution is **hypertonic.** Because the net movement of water would be out of the cell, the cell would *shrink* in a hypertonic solution. Conversely, if the concentration of impermeable solutes is *lower* outside of the cell than in the cell's interior, then the solution is **hypotonic.** The net movement of water would be into the cell, and the cell would *swell* and possibly burst.

EQUIPMENT USED The following equipment will be depicted on-screen: left and right beakers—used for diffusion of solutes; dialysis membranes with various molecular weight cutoffs (MWCOs).

Experiment Instructions

Go to the home page in the PhysioEx software and click **Exercise 1: Cell Transport Mechanisms and Permeability.** Click **Activity 3: Simulating Osmotic Pressure,** and take the online **Pre-lab Quiz** for Activity 3.

After you take the online Pre-lab Quiz, click the **Experiment** tab and begin the experiment. The experiment instructions are reprinted here for your reference. The opening screen for the experiment is shown below.

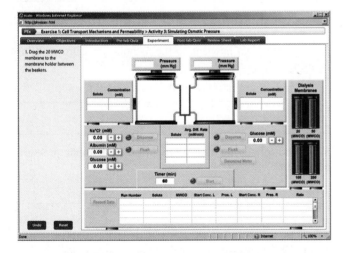

1. Drag the 20 MWCO membrane to the membrane holder between the beakers.

2. Increase the Na^+Cl^- concentration to be dispensed to the left beaker to 5.00 mM by clicking the + button beside the Na^+Cl^- display. Click **Dispense** to fill the left beaker with 5.00 mM Na^+Cl^- solution.

3. Note that the concentration of Na^+Cl^- in the left beaker is displayed in the concentration window to the left of the beaker. Click **Deionized Water** and then click **Dispense** to fill the right beaker with deionized water.

4. After you start the run, the barrier between the beakers will descend, allowing the solutions in each beaker to have access to the dialysis membrane separating them. You can observe the changes in pressure in the two beakers by watching the pressure display above each beaker. You will also be able to determine the amount of solute that passes through the membrane by observing the concentration display to the side of each beaker. A level above zero in Na^+Cl^- concentration in the right beaker indicates that Na^+ and Cl^- ions are diffusing from the left beaker into the right beaker through the selectively permeable dialysis membrane. Note that the timer is set to 60 minutes. The simulation compresses the 60-minute time period into 10 seconds of real time. Click **Start** to start the run and watch the pressure display above each beaker for any activity.

5. Click **Record Data** to display your results in the grid (and record your results in Chart 3).

CHART 3	Osmosis Results		
Solute	Membrane (MWCO)	Pressure on left (mm Hg)	Diffusion rate (mM/sec)
Na^+Cl^-			
Na^+Cl^-			
Na^+Cl^-			
Glucose			
Glucose			
Glucose			
Albumin w/glucose			

6. Click **Flush** beneath each of the beakers to prepare for the next run.

7. Increase the Na^+Cl^- concentration to be dispensed to the left beaker to 10.00 mM by clicking the + button beside the Na^+Cl^- display. Click **Dispense** to fill the left beaker with 10.00 mM Na^+Cl^- solution.

8. Click **Deionized Water** and then click **Dispense** to fill the right beaker with deionized water.

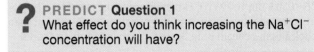

PREDICT Question 1
What effect do you think increasing the Na^+Cl^- concentration will have?

9. Click **Start** to start the run and watch the pressure display above each beaker for any activity.

10. Click **Record Data** to display your results in the grid (and record your results in Chart 3).

11. Click the 20 MWCO membrane in the membrane holder to automatically return it to the membrane cabinet and then click **Flush** beneath each beaker to prepare for the next run.

12. Drag the 50 MWCO membrane to the membrane holder between the beakers. Increase the Na^+Cl^- concentration to be dispensed to the left beaker to 10.00 mM. Click **Dispense** to fill the left beaker with 10.00 mM Na^+Cl^- solution.

13. Click **Deionized Water** and then click **Dispense** to fill the right beaker with deionized water.

14. Click **Start** to start the run and watch the pressure display above each beaker for any activity.

15. Click **Record Data** to display your results in the grid (and record your results in Chart 3).

16. Click the 50 MWCO membrane in the membrane holder to automatically return it to the membrane cabinet and then click **Flush** beneath each beaker to prepare for the next run.

17. Drag the 100 MWCO membrane to the membrane holder between the beakers. Increase the glucose concentration to be dispensed to the left beaker to 8.00 mM by clicking the + button beside the glucose display beneath the left beaker. Click **Dispense** to fill the left beaker with 8.00 mM glucose solution.

18. Click **Deionized Water** and then click **Dispense** to fill the right beaker with deionized water.

19. Click **Start** to start the run and watch the pressure display above each beaker for any activity.

20. Click **Record Data** to display your results in the grid (and record your results in Chart 3).

21. Click **Flush** beneath each of the beakers to prepare for the next run.

22. Increase the glucose concentration to be dispensed to the left beaker to 8.00 mM. Click **Dispense** to fill the left beaker with 8.00 mM glucose solution.

23. Increase the glucose concentration to be dispensed to the right beaker to 8.00 mM by clicking the + button beside the glucose display beneath the right beaker. Click **Dispense** to fill the right beaker with 8.00 mM glucose solution.

24. Click **Start** to start the run and watch the pressure display above each beaker for any activity.

25. Click **Record Data** to display your results in the grid (and record your results in Chart 3).

26. Click the 100 MWCO membrane in the membrane holder to automatically return it to the membrane cabinet and then click **Flush** beneath each beaker to prepare for the next run.

27. Drag the 200 MWCO membrane to the membrane holder between the beakers. Increase the glucose concentration to be dispensed to the left beaker to 8.00 mM. Click **Dispense** to fill the left beaker with 8.00 mM glucose solution.

28. Click **Deionized Water** and then click **Dispense** to fill the right beaker with deionized water.

29. Click **Start** to start the run and watch the pressure display above each beaker for any activity.

30. Click **Record Data** to display your results in the grid (and record your results in Chart 3).

31. Click **Flush** beneath each of the beakers to prepare for the next run.

32. Increase the albumin concentration to be dispensed to the left beaker to 9.00 mM. Click **Dispense** to fill the left beaker with 9.00 mM albumin solution.

33. Increase the glucose concentration to be dispensed to right beaker to 10.00 mM. Click **Dispense** to fill the right beaker with 10.00 mM glucose solution.

 PREDICT Question 2
What do you think will be the pressure result of the current experimental conditions?

34. Click **Start** to start the run and watch the pressure display above each beaker for any activity.

35. Click **Record Data** to display your results in the grid (and record your results in Chart 3).

After you complete the experiment, take the online **Post-lab Quiz** for Activity 3.

Activity Questions

1. Which membrane resulted in the greatest pressure with Na^+Cl^- as the solute? Why?

2. Explain what happens to the osmotic pressure with increasing solute concentration.

3. If the solutes are allowed to diffuse, is osmotic pressure generated?

4. If the solute concentrations are equal, is osmotic pressure generated? Why or why not?

Simulating Filtration

OBJECTIVES

1. To understand that filtration is a passive process dependent upon a pressure gradient.
2. To understand that filtration is not a selective process.
3. To explain that the size of the membrane pores will determine what passes through.
4. To explain the effect that increasing the hydrostatic pressure has on the filtration rate and how this correlates to events in the body.
5. To understand the relationship between molecular weight and molecular size.

Introduction

Filtration is the process by which water and solutes pass through a membrane (such as a dialysis membrane) from an area of higher hydrostatic (fluid) pressure into an area of lower hydrostatic pressure. Like diffusion, filtration is a passive process. For example, fluids and solutes filter out of the capillaries in the kidneys into the kidney tubules because blood pressure in the capillaries is greater than the fluid pressure in the tubules. So, if blood pressure increases, the rate of filtration increases.

Filtration is not a selective process. The amount of *filtrate*—the fluids and solutes that pass through the membrane—depends almost entirely on the *pressure gradient* (the difference in pressure between the solutions on the two sides of the membrane) and on the *size* of the *membrane pores*. Solutes that are too large to pass through are retained by the capillaries. These solutes usually include blood cells and proteins. Ions and smaller molecules, such as glucose and urea, can pass through.

In this activity the pore size is measured as a *molecular weight cutoff (MWCO),* which is indicated by the number below the filtration membrane. You can think of MWCO in terms of pore size: the larger the MWCO number, the larger the pores in the filtration membrane. The molecular weight of a solute is the number of grams per mole, where a mole is the constant Avogadro's number 6.02×10^{23} molecules/mole. You will also analyze the filtration membrane for the presence or absence of solutes that might be left sticking to the membrane.

> **EQUIPMENT USED** The following equipment will be depicted on-screen: top and bottom beakers—used for filtration of solutes; dialysis membranes with various molecular weight cutoffs (MWCOs); membrane residue analysis station—used to analyze the filtration membrane.

Experiment Instructions

Go to the home page in the PhysioEx software and click **Exercise 1: Cell Transport Mechanisms and Permeability.** Click **Activity 4: Simulating Filtration** and take the online **Pre-lab Quiz** for Activity 4.

After you take the online Pre-lab Quiz, click the **Experiment** tab and begin the experiment. The experiment instructions are reprinted here for your reference. The opening screen for the experiment is shown above.

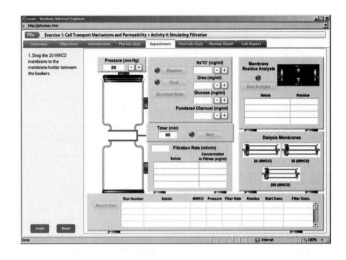

1. Drag the 20 MWCO membrane to the membrane holder between the beakers.

2. Increase the concentration of Na^+Cl^-, urea, glucose, and powdered charcoal to be dispensed to 5.00 mg/ml by clicking the + button beside the display for each solute. Click **Dispense** to fill the top beaker.

3. After you start the run, the membrane holder below the top beaker retracts, and the solution will filter through the membrane into the beaker below. You will be able to determine whether solute particles are moving through the filtration membrane by observing the concentration displays beside the bottom beaker. A rise in detected solute concentration indicates that the solute particles are moving through the filtration membrane. Note that the pressure is set at 50 mm Hg and the timer is set to 60 minutes. The simulation compresses the 60-minute time period into 10 seconds of real time. Click **Start** to start the run and watch the concentration displays beside the bottom beaker for any activity.

4. Drag the 20 MWCO membrane to the holder in the membrane residue analysis unit. Click **Start Analysis** to begin analysis (and cleaning) of the membrane.

5. Click **Record Data** to display your results in the grid (and record your results in Chart 4).

6. Click the 20 MWCO membrane in the membrane holder to automatically return it to the membrane cabinet and then click **Flush** to prepare for the next run.

> **PREDICT Question 1**
> What effect will increasing the pore size of the filter have on the filtration rate?

7. Drag the 50 MWCO membrane to the membrane holder between the beakers. With the concentration of Na^+Cl^-, urea, glucose, and powdered charcoal still set to 5.00 mg/ml, click **Dispense** to fill the top beaker.

8. Click **Start** to start the run and watch the concentration displays beside the bottom beaker for any activity.

CHART 4	Filtration Results				
		Membrane (MWCO)			
		20	50	200	200
Solute	Filtration rate (ml/sec)				
Na$^+$Cl$^-$	Filter concentration (mg/ml)				
	Membrane residue				
Urea	Filter concentration (mg/ml)				
	Membrane residue				
Glucose	Filter concentration (mg/ml)				
	Membrane residue				
Powdered charcoal	Filter concentration (mg/ml)				
	Membrane residue				

9. Drag the 50 MWCO membrane to the holder in the membrane residue analysis unit. Click **Start Analysis** to begin analysis (and cleaning) of the membrane.

10. Click **Record Data** to display your results in the grid (and record your results in Chart 4).

11. Click the 50 MWCO membrane in the membrane holder to automatically return it to the membrane cabinet and then click **Flush** to prepare for the next run.

12. Drag the 200 MWCO membrane to the membrane holder between the beakers. With the concentration of Na$^+$Cl$^-$, urea, glucose, and powdered charcoal still set to 5.00 mg/ml, click **Dispense** to fill the top beaker.

13. Click **Start** to start the run and watch the concentration displays beside the bottom beaker for any activity.

14. Drag the 200 MWCO membrane to the holder in the membrane residue analysis unit. Click **Start Analysis** to begin analysis (and cleaning) of the membrane.

15. Click **Record Data** to display your results in the grid (and record your results in Chart 4).

16. Click the 200 MWCO membrane in the membrane holder to automatically return it to the membrane cabinet and then click **Flush** to prepare for the next run.

> **? PREDICT Question 2**
> What will happen if you increase the pressure above the beaker (the driving pressure)?

17. Increase the pressure to 100 mm Hg by clicking on the + button beside the pressure display above the top beaker.

18. Drag the 200 MWCO membrane to the membrane holder between the beakers. With the concentration of Na$^+$Cl$^-$, urea, glucose, and powdered charcoal still set to 5.00 mg/ml, click **Dispense** to fill the top beaker.

19. Click **Start** to start the run and watch the concentration displays beside the bottom beaker for any activity.

20. Drag the 200 MWCO membrane to the holder in the membrane residue analysis unit. Click **Start Analysis** to begin analysis (and cleaning) of the membrane.

21. Click **Record Data** to display your results in the grid (and record your results in Chart 4).

After you complete the Experiment, take the online **Post-lab Quiz** for Activity 4.

Activity Questions

1. Explain your results with the 20 MWCO filter. Why weren't any of the solutes present in the filtrate?

_____

_____

2. Describe two variables that affected the rate of filtration in your experiments.

_____

_____

3. Explain how you can increase the filtration rate through living membranes.

_____

_____

4. Judging from the filtration results, indicate which solute has the largest molecular weight.

_____

ACTIVITY 5

Simulating Active Transport

OBJECTIVES

1. To understand that active transport requires cellular energy in the form of ATP.
2. To explain how the balance of sodium and potassium is maintained by the Na^+-K^+ pump, which moves both ions against their concentration gradients.
3. To understand coupled transport and be able to explain how the movement of sodium and potassium is independent of other solutes, such as glucose.

Introduction

Whenever a cell uses cellular energy (ATP) to move substances across its membrane, the process is an *active transport process.* Substances moved across cell membranes by an active transport process are generally unable to pass by diffusion. There are several reasons why a substance might not be able to pass through a membrane by diffusion: it might be too large to pass through the membrane pores, it might not be lipid soluble, or it might have to move *against*, rather than with, a concentration gradient.

In one type of active transport, substances move across the membrane by combining with a carrier-protein molecule. This kind of process resembles an enzyme-substrate interaction. ATP hydrolysis provides the driving force, and, in many cases, the substances move *against* concentration gradients or electrochemical gradients or both. The carrier proteins are commonly called **solute pumps.** Substances that are moved into cells by solute pumps include amino acids and some sugars. Both of these kinds of solutes are necessary for the life of the cell, but they are lipid insoluble and too large to pass through membrane pores.

In contrast, sodium ions (Na^+) are ejected from the cells by active transport. There is more Na^+ outside the cell than inside the cell, so Na^+ tends to remain in the cell unless actively transported out. In the body, the most common type of solute pump is the Na^+-K^+ (sodium-potassium) pump, which moves Na^+ and K^+ in opposite directions across cellular membranes. Three Na^+ ions are ejected from the cell for every two K^+ ions entering the cell. Note that there is more K^+ inside the cell than outside the cell, so K^+ tends to remain outside the cell unless actively transported in.

Membrane carrier proteins that move more than one substance, such as the Na^+-K^+ pump, participate in *coupled transport.* If the solutes move in the same direction, the carrier is a *symporter.* If the solutes move in opposite directions, the carrier is an *antiporter.* A carrier that transports only a single solute is a *uniporter.*

> **EQUIPMENT USED** The following equipment will be depicted on-screen: Simulated cell inside a large beaker.

Experiment Instructions

Go to the home page in the PhysioEx software and click **Exercise 1: Cell Transport Mechanisms and Permeability.** Click **Activity 5: Simulating Active Transport** and take the online **Pre-lab Quiz** for Activity 5.

After you take the online Pre-lab Quiz, click the **Experiment** tab and begin the experiment. The experiment instructions are reprinted here for your reference. The opening screen for the experiment is shown below.

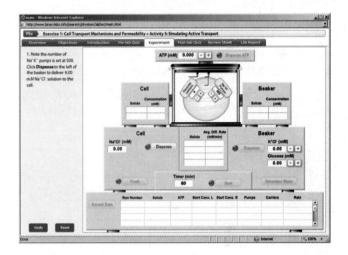

1. Note the number of Na^+-K^+ pumps is set at 500. Click **Dispense** to the left of the beaker to deliver 9.00 mM Na^+Cl^- solution to the cell.

2. Increase the K^+Cl^- concentration to be delivered to the beaker to 6.00 mM by clicking the + button beside the K^+Cl^- display. Click **Dispense** to the right of the beaker to deliver 6.00 mM K^+Cl^- solution to the beaker.

3. Increase the ATP concentration to 1.00 mM by clicking the + button beside the ATP display above the beaker. Click **Dispense ATP** to deliver 1.00 mM ATP solution to both sides of the membrane.

4. After you start the run, the solutes will move across the cell membrane, simulating active transport. You will be able to determine the amount of solute that is transported across the membrane by observing the concentration displays on both sides of the beaker (the display on the left shows the concentrations inside the cell and the display on the right shows the concentrations inside the beaker). Note that the timer is set to 60 minutes. The simulation compresses the 60-minute time period into 10 seconds of real time. Click **Start** to start the run and watch the concentration displays on both sides of the beaker for any activity.

5. Click **Record Data** to display your results in the grid.

6. Click **Flush** to reset the beaker and simulated cell.

7. Click **Dispense** to the left of the beaker to deliver 9.00 mM Na$^+$Cl$^-$ solution to the cell.

8. Increase the K$^+$Cl$^-$ concentration to be delivered to the beaker to 6.00 mM by clicking the + button beside the K$^+$Cl$^-$ display. Click **Dispense** to the right of the beaker to deliver 6.00 mM K$^+$Cl$^-$ solution to the beaker.

9. Increase the ATP concentration to 3.00 mM by clicking the + button beside the ATP display above the beaker. Click **Dispense ATP** to deliver 3.00 mM ATP solution to both sides of the membrane.

10. Click **Start** to start the run and watch the concentration displays on both sides of the beaker for any activity.

11. Click **Record Data** to display your results in the grid.

12. Click **Flush** to reset the beaker and simulated cell.

13. Click **Dispense** to the left of the beaker to deliver 9.00 mM Na$^+$Cl$^-$ solution to the cell.

14. Click **Deionized Water** to the right of the beaker and then click **Dispense** to deliver deionized water to the beaker.

15. Increase the ATP concentration to 3.00 mM. Click **Dispense ATP** to deliver 3.00 mM ATP solution to both sides of the membrane.

? PREDICT Question 1
What do you think will result from these experimental conditions?

_____

16. Click **Start** to start the run and watch the concentration displays on both sides of the beaker for any activity.

17. Click **Record Data** to display your results in the grid.

18. Click **Flush** to reset the beaker and simulated cell.

19. Increase the number of Na$^+$-K$^+$ pumps to 800 by clicking the + button beneath the Na$^+$-K$^+$ pump display. Click **Dispense** to the left of the beaker to deliver 9.00 mM Na$^+$Cl$^-$ solution to the cell.

20. Increase the K$^+$Cl$^-$ concentration to be delivered to the beaker to 6.00 mM. Click **Dispense** to the right of the beaker to deliver 6.00 mM K$^+$Cl$^-$ solution to the beaker.

21. Increase the ATP concentration to 3.00 mM. Click **Dispense ATP** to deliver 3.00 mM ATP solution to both sides of the membrane.

22. Click **Start** to start the run and watch the concentration displays on both sides of the beaker for any activity.

23. Click **Record Data** to display your results in the grid.

24. Click **Flush** to reset the beaker and simulated cell.

25. With the number of Na$^+$-K$^+$ pumps still set to 800, increase the number of glucose carriers to 400 by clicking the + button beneath the glucose carriers display. Click **Dispense** to the left of the beaker to deliver 9.00 mM Na$^+$Cl$^-$ solution to the cell.

? PREDICT Question 2
Do you think the addition of glucose carriers will affect the transport of sodium or potassium?

_____

26. Increase the K$^+$Cl$^-$ concentration to be delivered to the beaker to 6.00 mM. Increase the glucose concentration to be delivered to the beaker to 10.00 mM. Click **Dispense** to the right of the beaker to deliver 6.00 mM K$^+$Cl$^-$ and 10.00 mM glucose solution to the beaker.

27. Increase the ATP concentration to 3.00 mM. Click **Dispense ATP** to deliver 3.00 mM ATP solution to both sides of the membrane.

28. Click **Start** to start the run and watch the concentration displays on both sides of the beaker for any activity.

29. Click **Record Data** to display your results in the grid.

After you complete the experiment, take the online **Post-lab Quiz** for Activity 5.

Activity Questions

1. In the initial trial the number of Na$^+$-K$^+$ pumps is set to 500, the Na$^+$Cl$^-$ concentration is set to 9.00 mM, the K$^+$Cl$^-$ concentration is set to 6.00 mM, and the ATP concentration is set to 1.00 mM. Explain what happened and why. What would happen if no ATP had been dispensed?

_____

_____

2. Why was there no transport when you dispensed only Na$^+$Cl$^-$, even though ATP was present?

_____

_____

3. What happens to the rate of transport of Na$^+$ and K$^+$ when you increase the number of Na$^+$-K$^+$ pumps?

_____

_____

4. Explain why the Na$^+$ and K$^+$ transports were unaffected by the addition of glucose.

_____

_____

Cell Transport Mechanisms and Permeability

NAME_____

LAB TIME/DATE_____

A C T I V I T Y   1 Simulating Dialysis (Simple Diffusion)

1. Describe two variables that affect the rate of diffusion. _____

_____

2. Why do you think the urea was not able to diffuse through the 20 MWCO membrane? How well did the results compare with

your prediction? _____

3. Describe the results of the attempts to diffuse glucose and albumin through the 200 MWCO membrane. How well did the

results compare with your prediction? _____

_____

4. Put the following in order from smallest to largest molecular weight: glucose, sodium chloride, albumin, and urea. _____

_____

A C T I V I T Y   2 Simulated Facilitated Diffusion

1. Explain one way in which facilitated diffusion is the same as simple diffusion and one way in which it differs. _____

_____

2. The larger value obtained when more glucose carriers were present corresponds to an increase in the rate of glucose

transport. Explain why the rate increased. How well did the results compare with your prediction? _____

_____

3. Explain your prediction for the effect Na^+Cl^- might have on glucose transport. In other words, explain why you picked the

choice that you did. How well did the results compare with your prediction? _____

A C T I V I T Y   3 Simulating Osmotic Pressure

1. Explain the effect that increasing the Na^+Cl^- concentration had on osmotic pressure and why it has this effect. How well did

the results compare with your prediction? _____

_____

2. Describe one way in which osmosis is similar to simple diffusion and one way in which it is different. _____

    _____

3. Solutes are sometimes measured in milliosmoles. Explain the statement, "Water chases milliosmoles." _____

    _____

4. The conditions were 9 mM albumin in the left beaker and 10 mM glucose in the right beaker with the 200 MWCO membrane in place. Explain the results. How well did the results compare with your prediction? _____

    _____

ACTIVITY 4   Simulating Filtration

1. Explain in your own words why increasing the pore size increased the filtration rate. Use an analogy to support your statement. How well did the results compare with your prediction? _____

    _____

2. Which solute did not appear in the filtrate using any of the membranes? Explain why. _____

    _____

3. Why did increasing the pressure increase the filtration rate but not the concentration of solutes? How well did the results compare with your prediction? _____

    _____

ACTIVITY 5   Simulating Active Transport

1. Describe the significance of using 9 mM sodium chloride inside the cell and 6 mM potassium chloride outside the cell, instead of other concentration ratios. _____

    _____

2. Explain why there was no sodium transport even though ATP was present. How well did the results compare with your prediction? _____

3. Explain why the addition of glucose carriers had no effect on sodium or potassium transport. How well did the results compare with your prediction? _____

4. Do you think glucose is being actively transported or transported by facilitated diffusion in this experiment? Explain your answer. _____

    _____

Skeletal Muscle Physiology

Exercise Overview

Humans make voluntary decisions to walk, talk, stand up, and sit down. Skeletal muscles, which are usually attached to the skeleton, make these actions possible (view Figure 2.1). Skeletal muscles characteristically span two joints and attach to the skeleton via **tendons,** which attach to the periosteum of a bone. Skeletal muscles are composed of hundreds to thousands of individual cells called **muscle fibers,** which produce **muscle tension** (also referred to as **muscle force**). Skeletal muscles are remarkable machines. They provide us with the manual dexterity to create magnificent works of art and can generate the brute force needed to lift a 45-kilogram sack of concrete.

When a skeletal muscle is isolated from an experimental animal and mounted on a **force transducer,** you can generate **muscle contractions** with controlled **electrical stimulation.** Importantly, the contractions of this isolated muscle are known to mimic those of working muscles in the body. That is, in vitro experiments reproduce in vivo functions. Therefore, the activities you perform in this exercise will give you valuable insight into skeletal muscle physiology.

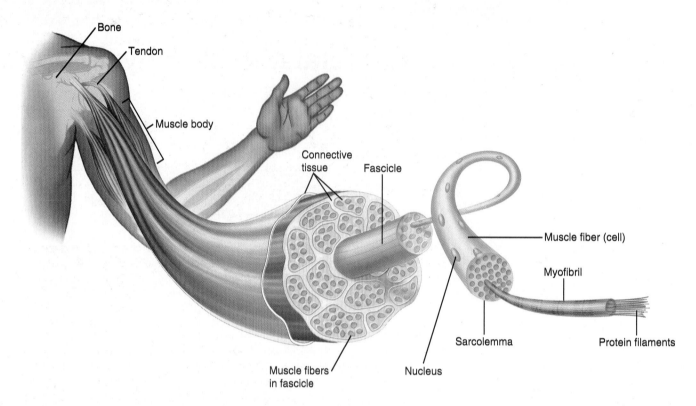

FIGURE 2.1 Structure of a skeletal muscle.

The Muscle Twitch and the Latent Period

OBJECTIVES

1. To understand the terms *excitation-contraction coupling, electrical stimulus, muscle twitch, latent period, contraction phase,* and *relaxation phase.*
2. To initiate muscle twitches with electrical stimuli of varying intensity.
3. To identify and measure the duration of the latent period.

Introduction

A **motor unit** consists of a **motor neuron** and all of the **muscle fibers** it innervates. The motor neuron and a muscle fiber intersect at the **neuromuscular junction.** Specifically, the neuromuscular junction is the location where the axon terminal of the neuron meets a specialized region of the muscle fiber's plasma membrane. This specialized region is called the **motor end plate.**

An action potential in a motor neuron triggers the release of acetylcholine from its terminal. Acetylcholine then diffuses onto the muscle fiber's plasma membrane (or **sarcolemma**) and binds to receptors in the motor end plate, initiating a change in ion permeability that results in a *graded depolarization* of the muscle plasma membrane (the end-plate potential). The events that occur at the neuromuscular junction lead to the **end-plate potential.** The end-plate potential triggers a series of events that results in the contraction of a muscle cell. This entire process is called **excitation-contraction coupling.**

You will be simulating excitation-contraction coupling in this and subsequent activities, but you will be using electrical pulses, rather than acetylcholine, to trigger action potentials. The pulses will be administered by an electrical stimulator that can be set for the precise voltage, frequency, and duration of shock desired. When applied to a muscle that has been surgically removed from an animal, a single electrical stimulus will result in a **muscle twitch**—the mechanical response to a single action potential. A muscle twitch has three phases: the *latent period,* the *contraction phase,* and the *relaxation phase.*

1. The **latent period** is the period of time that elapses between the generation of an action potential in a muscle cell and the start of muscle contraction. Although no force is generated during the latent period, chemical changes (including the release of calcium from the sarcoplasmic reticulum) occur intracellularly in preparation for contraction.

2. The **contraction phase** starts at the end of the latent period and ends when muscle tension peaks.

3. The **relaxation phase** is the period of time from peak tension until the end of the muscle contraction.

> **EQUIPMENT USED** The following equipment will be depicted on-screen: intact, viable skeletal muscle dissected off the leg of a frog; electrical stimulator—delivers the desired amount and duration of stimulating voltage to the muscle via electrodes resting on the muscle; mounting stand—includes a force transducer to measure the amount of force, or tension, developed by the muscle; oscilloscope—displays the stimulated muscle twitch and the amount of active, passive, and total force developed by the muscle.

Experiment Instructions

Go to the home page in the PhysioEx software and click **Exercise 2: Skeletal Muscle Physiology.** Click **Activity 1: The Muscle Twitch and the Latent Period,** and take the online **Pre-lab Quiz** for Activity 1.

After you take the online Pre-lab Quiz, click the **Experiment** tab and begin the experiment. The experiment instructions are reprinted here for your reference. The opening screen for the experiment is shown below.

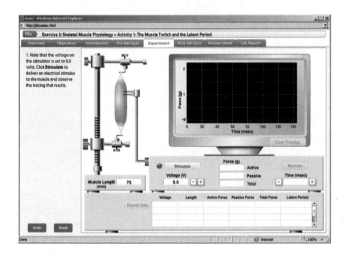

1. Note that the voltage on the stimulator is set to 0.0 volts. Click **Stimulate** to deliver an electrical stimulus to the muscle and observe the tracing that results.

2. The tracing on the oscilloscope indicates active muscle force. Note that no muscle force developed because the voltage was set to zero. Click **Record Data** to display your results in the grid (and record your results in Chart 1).

CHART 1	Latent Period Results	
Voltage	Active force (g)	Latent period (msec)

3. Increase the voltage to 3.0 volts by clicking the + button beside the voltage display.

4. Click **Stimulate** and observe the tracing that results.

5. Note the muscle force that developed. Click **Record Data** to display your results in the grid (and record your results in Chart 1).

6. Click **Clear Tracings** to remove the tracings from the oscilloscope.

7. Increase the voltage to 4.0 volts by clicking the + button beside the voltage display.

8. Click **Stimulate** and observe the tracing that results. Note that the trace starts at the left side of the screen and stays flat for a short period of time. Remember that the X-axis displays elapsed time in milliseconds. Also note how the force during the twitch also changes.

9. Click **Measure** on the stimulator. A thin, vertical yellow line appears at the far left side of the oscilloscope screen. To measure the length of the latent period, you measure the time between the application of the stimulus and the beginning of the first observable response (here, an increase in force). Click the + button beside the time display. You will see the vertical yellow line start to move across the screen. Watch what happens in the time (msec) display as the line moves across the screen. Keep clicking the + button until the yellow line reaches the point in the tracing where the graph stops being a flat line and begins to rise (this is the point at which muscle tension starts to develop). If the yellow line moves past the desired point, click the − button to move it backward.

When the yellow line is positioned correctly, click **Record Data** to display the latent period in the grid (and record your results in Chart 1).

10. Click **Clear Tracings** to remove the tracings from the oscilloscope.

 PREDICT **Question 1**
Will changes to the stimulus voltage alter the duration of the latent period? Explain.

11. You will now gradually increase the voltage to observe how changes to the stimulus voltage alter the duration of the latent period.

- Increase the voltage by 2.0 volts.
- Click **Stimulate** and observe the tracing that results.
- Click **Measure** on the stimulator and then click the + button until the yellow line reaches the point in the tracing where the graph stops being a flat line and begins to rise.
- Click **Record Data** (and record your results in Chart 1).

Repeat this step until you reach 10.0 volts.

After you complete the experiment, take the online **Post-lab Quiz** for Activity 1.

Activity Questions

1. Draw a graph that depicts a single skeletal muscle twitch, placing time on the X-axis and force on the Y-axis. Label the phases of this muscle twitch and describe what is happening in the muscle during each phase.

2. During the latent period of a skeletal muscle twitch, there is an apparent lack of muscle activity. Describe the electrical and chemical changes that occur in the muscle during this period.

_____

_____

_____ ▬

The Effect of Stimulus Voltage on Skeletal Muscle Contraction

OBJECTIVES

1. To understand the terms *motor neuron, muscle twitch, motor unit, recruitment, stimulus voltage, threshold stimulus,* and *maximal stimulus.*
2. To understand how motor unit recruitment can increase the tension a whole muscle develops.
3. To identify a threshold stimulus voltage.
4. To observe the effect of increases in stimulus voltage on a whole muscle.
5. To understand how increasing stimulus voltage to an isolated muscle in an experiment mimics motor unit recruitment in the body.

Introduction

A skeletal muscle produces **tension** (also known as **muscle force**) when nervous or electrical stimulation is applied. The force generated by a whole muscle reflects the number of active **motor units** at a given moment. A strong muscle contraction implies that many motor units are activated, with each unit developing its maximal tension, or force. A weak muscle contraction implies that fewer motor units are activated, but each motor unit still develops its maximal tension. By increasing the number of active motor units, we can produce a steady increase in muscle force, a process called **motor unit recruitment.**

Regardless of the number of **motor units** activated, a single stimulated contraction of whole skeletal muscle is called a **muscle twitch.** A tracing of a muscle twitch is divided into three phases: the latent period, the contraction phase, and the relaxation phase. The latent period is a short period between the time of muscle stimulation and the beginning of a muscle response. Although no force is generated during this interval, chemical changes occur intracellularly in preparation for contraction (including the release of calcium from the sarcoplasmic reticulum). During the contraction phase, the myofilaments utilize the cross-bridge cycle and the muscle develops tension. Relaxation takes place when the contraction has ended and the muscle returns to its normal resting state and length.

In this activity you will stimulate an isometric, or fixed-length, contraction of an isolated skeletal muscle. This activity allows you to investigate how the strength of an electrical stimulus affects whole-muscle function. Note that these simulations involve indirect stimulation by an electrode placed on the surface of the muscle. Indirect stimulation differs from the situation in vivo, where each fiber in the muscle receives direct stimulation via a nerve ending. Nevertheless, increasing the intensity of the electrical stimulation mimics how the nervous system increases the number of activated motor units.

The **threshold voltage** is the smallest stimulus required to induce an action potential in a muscle fiber's plasma membrane, or sarcolemma. As the **stimulus voltage** to a muscle is increased beyond the threshold voltage, the amount of force produced by the whole muscle also increases. This result occurs because, as more voltage is delivered to the whole muscle, more muscle fibers are activated and, thus, the total force produced by the muscle increases. Maximal tension in the whole muscle occurs when all the muscle fibers have been activated by a sufficiently strong stimulus (referred to as the **maximal voltage**). Stimulation with voltages greater than the maximal voltage will not increase the force of contraction. This experiment is analogous to, and accurately mimics, muscle activity in vivo, where the recruitment of additional motor units increases the total muscle force produced. This phenomenon is called *motor unit recruitment*.

> **EQUIPMENT USED** The following equipment will be depicted on-screen: intact, viable skeletal muscle dissected off the leg of a frog; electrical stimulator—delivers the desired amount and duration of stimulating voltage to the muscle via electrodes resting on the muscle; mounting stand—includes a force transducer to measure the amount of force, or tension, developed by the muscle; oscilloscope—displays the stimulated muscle twitch and the amount of active, passive, and total force developed by the muscle.

Experiment Instructions

Go to the home page in the PhysioEx software and click **Exercise 2: Skeletal Muscle Physiology.** Click **Activity 2: The Effect of Stimulus Voltage on Skeletal Muscle Contraction,** and take the online **Pre-lab Quiz** for Activity 2.

After you take the online Pre-lab Quiz, click the **Experiment** tab and begin the experiment. The experiment instructions are reprinted here for your reference. The opening screen for the experiment is shown on the following page.

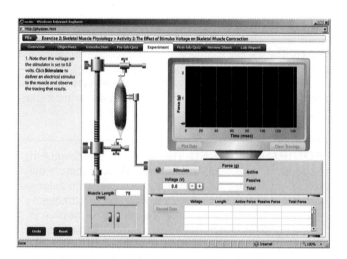

1. Note that the voltage on the stimulator is set to 0.0 volts. Click **Stimulate** to deliver an electrical stimulus to the muscle and observe the tracing that results.

2. Note the active force display and then click **Record Data** to display your results in the grid (and record your results in Chart 2).

3. Increase the voltage to 0.2 volts by clicking the + button beside the voltage display. Click **Stimulate** to deliver an electrical stimulus to the muscle and observe the tracing that results.

4. Note the active force display and then click **Record Data** to display your results in the grid (and record your results in Chart 2).

5. You will now gradually increase the voltage and stimulate the muscle to determine the minimum voltage required to generate active force.

  • Increase the voltage by 0.1 volts and then click **Stimulate.**

  • If no active force is generated, increase the voltage by 0.1 volts and stimulate the muscle again. When active force is generated, click **Record Data** to display your results in the grid (and record your results in Chart 2).

6. Enter the threshold voltage for this experiment in the field below and then click **Submit** to record your answer in the lab report. _____ volts

7. Click **Clear Tracings** to clear the tracings on the oscilloscope.

? **PREDICT Question 1**
As the stimulus voltage is increased from 1.0 volt up to 10 volts, what will happen to the amount of active force generated with each stimulus?

_____

8. Increase the voltage on the stimulator to 1.0 volt and then click **Stimulate.**

9. Note the active force display and then click **Record Data** to display your results in the grid (and record your results in Chart 2).

CHART 2	Effect of Stimulus Voltage on Skeletal Muscle Contraction
Voltage	Active force (g)

10. You will now gradually increase the voltage and stimulate the muscle to determine the maximal voltage.

  • Increase the voltage by 0.5 volts.

  • Click **Stimulate** and observe the tracing that results.

  • Note the active force display and then click **Record Data** to display your results in the grid (and record your results in Chart 2).

  Repeat this step until you reach 10.0 volts.

11. Click **Plot Data** to view a summary of your data on a plotted grid. Click **Submit** to record your plot in the lab report.

12. Enter the maximal voltage for this experiment in the field below and then click **Submit** to record your answer in the lab report. _____ volts

After you complete the experiment, take the online **Post-lab Quiz** for Activity 2.

Activity Questions

1. For a single skeletal muscle twitch, explain the effect of increasing stimulus voltage.

_____

_____

2. How is this effect achieved in vivo?

_____

_____ ▬

The Effect of Stimulus Frequency on Skeletal Muscle Contraction

OBJECTIVES

1. To understand the terms *stimulus frequency, wave summation,* and *treppe.*
2. To observe the effect of an increasing stimulus frequency on the force developed by an isolated skeletal muscle.
3. To understand how increasing stimulus frequency to an isolated skeletal muscle induces the summation of twitch force.

Introduction

As demonstrated in Activity 2, increasing the stimulus voltage to an isolated skeletal muscle (up to a maximal value) results in an increase of force produced by the whole muscle. This experimental result is analogous to motor unit recruitment in the body. Importantly, this result relies on being able to increase the single stimulus intensity in the experiment. You will now explore another way to increase the force produced by an isolated skeletal muscle.

When a muscle first contracts, the force it is able to produce is less than the force it is able to produce with subsequent stimulations within a relatively short time span. **Treppe** is the progressive increase in force generated when a muscle is stimulated in succession, such that muscle twitches follow one another closely, with each successive twitch peaking slightly higher than the one before. This step-like increase in force is why treppe is also known as the staircase effect. For the first few twitches, each successive twitch produces slightly more force than the previous twitch as long as the muscle is allowed to fully relax between stimuli and the stimuli are delivered relatively close together.

When a skeletal muscle is stimulated repeatedly, such that the stimuli arrive one after another within a short period of time, muscle twitches can overlap with each other and result in a stronger muscle contraction than a stand-alone twitch. This phenomenon is known as wave summation.

Wave summation occurs when muscle fibers that are developing tension are stimulated again before the fibers have relaxed. Thus, wave summation is achieved by increasing the **stimulus frequency,** or rate of stimulus delivery to the muscle. Wave summation occurs because the muscle fibers are already in a partially contracted state when subsequent stimuli are delivered.

> **EQUIPMENT USED** The following equipment will be depicted on-screen: intact, viable skeletal muscle dissected off the leg of a frog; an electrical stimulator—delivers the desired amount and duration of stimulating voltage to the muscle via electrodes resting on the muscle; mounting stand—includes a force transducer to measure the amount of force, or tension, developed by the muscle; oscilloscope—displays the stimulated muscle twitch and the amount of active, passive, and total force developed by the muscle.

Experiment Instructions

Go to the home page in the PhysioEx software and click **Exercise 2: Skeletal Muscle Physiology.** Click **Activity 3: The Effect of Stimulus Frequency on Skeletal Muscle Contraction,** and take the online **Pre-lab Quiz** for Activity 3.

After you take the online Pre-lab Quiz, click the **Experiment** tab and begin the experiment. The experiment instructions are reprinted here for your reference. The opening screen for the experiment is shown below.

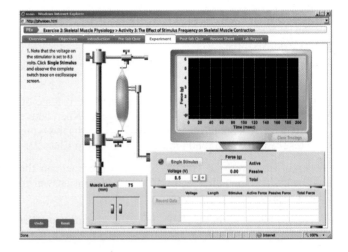

1. Note that the voltage on the stimulator is set to 8.5 volts. Click **Single Stimulus** and observe the tracing that results on the oscilloscope.

2. Note the active force display and then click **Record Data** to display your results in the grid (and record your results in Chart 3).

3. Click **Single Stimulus** and allow the trace to rise and completely fall. *Immediately after* the trace has returned to baseline, click **Single Stimulus** again.

CHART 3	Effect of Stimulus Frequency on Skeletal Muscle Contraction	
Voltage	Stimulus	Active force (g)

4. Note the active force for the second muscle twitch and click **Record Data** to display your results in the grid (and record your results in Chart 3).

5. You should have observed an increase in active force generated by the muscle with the immediate second stimulus. This increase demonstrates the phenomenon of treppe. Click **Clear Tracings** to clear the tracings on the oscilloscope.

6. You will now investigate the process of wave summation. Click **Single Stimulus** and watch the trace rise and begin to fall. *Before* the trace falls completely back to the baseline, click **Single Stimulus** again. (You can simply click **Single Stimulus** twice in quick succession in order to achieve this.)

7. Note the active force for the second muscle twitch and click **Record Data** to display your results in the grid (and record your results in Chart 3).

? PREDICT Question 1
As the stimulus frequency increases, what will happen to the muscle force generated with each successive stimulus? Will there be a limit to this response?

_____

_____

_____

8. Now stimulate the muscle at a higher frequency by clicking **Single Stimulus** four times in rapid succession.

9. Note the active force display and then click **Record Data** to display your results in the grid (and record your results in Chart 3).

10. Click **Clear Tracings** to clear the tracings on the oscilloscope.

? PREDICT Question 2
In order to produce sustained muscle contractions with an active force value of 5.2 grams, do you think you need to increase the stimulus voltage?

_____

11. Increase the voltage to 10.0 volts by clicking the **+** button beside the voltage display. After setting the voltage, click **Single Stimulus** four times in rapid succession.

12. Note the active force display and then click **Record Data** to display your results in the grid (and record your results in Chart 3).

13. Click **Clear Tracings** to clear the tracings on the oscilloscope.

14. Return the voltage to 8.5 volts by clicking the **−** button beside the voltage display. After setting the voltage, click **Single Stimulus** as many times as you can in rapid succession. Note the active force display. If you did not achieve an active force of 5.2 grams, click **Clear Tracings** and then click **Single Stimulus** even more rapidly. Repeat this step until you achieve an active force of 5.2 grams.

When you achieve an active force of 5.2 grams, click **Record Data** to display your results in the grid (and record your results in Chart 3).

After you complete the experiment, take the online **Post-lab Quiz** for Activity 3.

Activity Questions

1. Why is treppe also known as the staircase effect?

_____

_____

2. What changes are thought to occur in the skeletal muscle to allow treppe to be observed?

_____

_____

3. How does the frequency of stimulation affect the amount of force generated by a skeletal muscle?

_____

_____

4. Explain how wave summation is achieved in vivo.

_____

_____

Tetanus in Isolated Skeletal Muscle

OBJECTIVES

1. To understand the terms *stimulus frequency, unfused tetanus, fused tetanus,* and *maximal tetanic tension.*
2. To observe the effect of an increasing stimulus frequency on an isolated skeletal muscle.
3. To understand how increasing the stimulus frequency to an isolated skeletal muscle leads to unfused or fused tetanus.

Introduction

As demonstrated in Activity 3, increasing the **stimulus frequency** to an isolated skeletal muscle results in an increase in force produced by the whole muscle. Specifically, you observed that, if electrical stimuli are applied to a skeletal muscle in quick succession, the overlapping twitches generated more force with each successive stimulus. However, if stimuli continue to be applied frequently to a muscle over a prolonged period of time, the maximum possible muscle force from each stimulus will eventually reach a plateau—a state known as **unfused tetanus.** If stimuli are then applied with even greater frequency, the twitches will begin to fuse so that the peaks and valleys of each twitch become indistinguishable from one another—this state is known as **complete (fused) tetanus.** When the stimulus frequency reaches a value beyond which no further increases in force are generated by the muscle, the muscle has reached its **maximal tetanic tension.**

> **EQUIPMENT USED** The following equipment will be depicted on-screen: intact, viable skeletal muscle dissected off the leg of a frog; electrical stimulator—delivers the desired amount and duration of stimulating voltage to the muscle via electrodes resting on the muscle; mounting stand—includes a force transducer to measure the amount of force, or tension, developed by the muscle; oscilloscope—displays the stimulated muscle twitch and the amount of active, passive, and total force developed by the muscle.

Experiment Instructions

Go to the home page in the PhysioEx software and click **Exercise 2: Skeletal Muscle Physiology.** Click **Activity 4: Tetanus in Isolated Skeletal Muscle** and take the online **Pre-lab Quiz** for Activity 4.

After you take the online Pre-lab Quiz, click the **Experiment** tab and begin the experiment. The experiment instructions are reprinted here for your reference. The opening screen for the experiment is shown above.

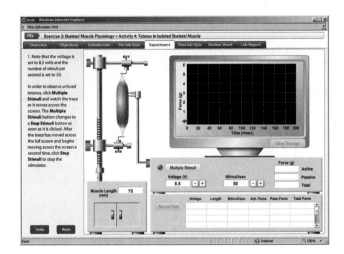

1. Note that the voltage is set to 8.5 volts and the number of stimuli per second is set to 50. To observe *unfused* tetanus, click **Multiple Stimuli** and watch the trace as it moves across the screen. The **Multiple Stimuli** button changes to a **Stop Stimuli** button after it is clicked. After the trace has moved across the full screen and begins moving across the screen a second time, click **Stop Stimuli** to stop the stimulator.

2. Click **Record Data** to display your results in the grid (and record your results in Chart 4).

CHART 4	Tetanus in Isolated Skeletal Muscle
Stimuli/second	Active force (g)

? PREDICT Question 1
As the stimulus frequency increases further, what will happen to the muscle tension and twitch appearance with each successive stimulus? Will there be a limit to this response?

_____

_____

3. In order to observe *fused* tetanus, increase the stimuli/sec setting to 130 by clicking the + button beside the stimuli/sec display. Click **Multiple Stimuli** and observe the resulting trace. After the trace has moved across the full screen and begins moving across the screen a second time, click **Stop Stimuli.**

4. Note the fused tetanus and click **Record Data** to display your results in the grid (and record your results in Chart 4).

5. Click **Clear Tracings** to clear the oscilloscope screen.

6. Increase the stimuli/sec setting to 140 by clicking the + button beside the stimuli/sec display. Click **Multiple Stimuli** and observe the resulting trace. After the trace has moved across the full screen and begins moving across the screen a second time, click **Stop Stimuli.**

7. Note the fused tetanus and click **Record Data** to display your results in the grid (and record your results in Chart 4).

8. Click **Clear Tracings** to clear the oscilloscope screen.

9. You will now observe the effect of incremental increases in the number of stimuli per second above 140 stimuli per second.

- Increase the stimuli/sec setting by 2.

- Click **Multiple Stimuli** and observe the resulting trace. After the trace has moved across the full screen and begins moving across the screen a second time, click **Stop Stimuli.**

- Click **Record Data** to display your results in the grid (and record your results in Chart 4).

- Click **Clear Tracings** to clear the oscilloscope screen.

Repeat this step until you reach 150 stimuli per second.

After you complete the experiment, take the online **Post-lab Quiz** for Activity 4.

Activity Questions

1. Explain what you think is being summated in the skeletal muscle to allow a high stimulus frequency to induce a smooth, continuous skeletal muscle contraction.

_____

_____

2. Why do many toddlers receive a tetanus shot (and then subsequent booster shots, as needed, later in life)? How does the condition known as "lockjaw" relate to tetanus shots?

_____

_____

_____

Fatigue in Isolated Skeletal Muscle

OBJECTIVES

1. To understand the terms *stimulus frequency, complete (fused) tetanus, fatigue,* and *rest period.*

2. To observe the development of skeletal muscle fatigue.

3. To understand how the length of intervening rest periods determines the onset of fatigue.

Introduction

As demonstrated in Activities 3 and 4, increasing the stimulus frequency to an isolated skeletal muscle induces an increase of force produced by the whole muscle. Specifically, if voltage stimuli are applied to a muscle frequently in quick succession, the skeletal muscle generates more force with each successive stimulus.

However, if stimuli continue to be applied frequently to a muscle over a prolonged period of time, the maximum force of each twitch eventually reaches a plateau—a state known as *unfused tetanus.* If stimuli are then applied with even greater frequency, the twitches begin to fuse so that the peaks and valleys of each twitch become indistinguishable from one another—this state is known as **complete (fused) tetanus.** When the **stimulus frequency** reaches a value beyond which no further increase in force is generated by the muscle, the muscle has reached its **maximal tetanic tension.**

In this activity you will observe the phenomena of skeletal muscle *fatigue.* Fatigue refers to a decline in a skeletal muscle's ability to maintain a constant level of force, or tension, after prolonged, repetitive stimulation. You will also demonstrate how intervening **rest periods** alter the onset of fatigue in skeletal muscle. The causes of fatigue are still being investigated and multiple molecular events are thought to be involved, though the accumulations of lactic acid, ADP, and P_i in muscles are thought to be the major factors causing fatigue in the case of high-intensity exercise.

Common definitions for **fatigue** are:

- The failure of a muscle fiber to produce tension because of previous contractile activity.

- A decline in the muscle's ability to maintain a constant force of contraction after prolonged, repetitive stimulation.

> **EQUIPMENT USED** The following equipment will be depicted on-screen: intact, viable skeletal muscle dissected off the leg of a frog; electrical stimulator—delivers the desired amount and duration of stimulating voltage to the muscle via electrodes resting on the muscle; mounting stand—includes a force transducer to measure the amount of force, or tension, developed by the muscle; oscilloscope—displays the stimulated muscle twitch and the amount of active, passive, and total force developed by the muscle.

Experiment Instructions

Go to the home page in the PhysioEx software and click **Exercise 2: Skeletal Muscle Physiology.** Click **Activity 5: Fatigue in Isolated Skeletal Muscle,** and take the online **Pre-lab Quiz** for Activity 5.

After you take the online Pre-lab Quiz, click the **Experiment** tab and begin the experiment. The experiment instructions are reprinted here for your reference. The opening screen for the experiment is shown on the following page.

1. Note that the voltage is set to 8.5 volts and the number of stimuli per second is set to 120. Click **Multiple Stimuli** and closely watch the muscle force tracing on the oscilloscope. Click **Stop Stimuli** after the muscle force falls to 0.

2. Click **Record Data** to display your results in the grid (and record your results in Chart 5).

CHART 5	Fatigue Results	
Rest period (sec)	Active force (g)	Sustained maximal force (sec)

3. Click **Clear Tracings** to clear the oscilloscope screen.

> **? PREDICT Question 1**
> If the stimulator is briefly turned off for defined periods of time, what will happen to the length of time that the muscle is able to sustain maximal developed tension when the stimulator is turned on again?
>
> _____
>
> _____

4. To demonstrate the onset of fatigue after a variable rest period, you will be clicking the **Multiple Stimuli** button on and off three times. Read through the steps below before proceeding. Watch the timer closely to help you determine when to turn the stimulator back on.

- Click **Multiple Stimuli.**
- After the muscle force falls to 0, click **Stop Stimuli** to turn off the stimulator.
- Wait 10 seconds, then click **Multiple Stimuli** to turn the stimulator back on.
- Click **Stop Stimuli** after the muscle force falls to 0.

- Wait 20 seconds, then click **Multiple Stimuli** to turn the stimulator back on.
- Click **Stop Stimuli** after the muscle force falls to 0.

5. Click **Record Data** to display your results in the grid (and record your results in Chart 5).

After you complete the experiment, take the online **Post-lab Quiz** for Activity 5.

Activity Questions

1. What proposed mechanisms most likely explain why fatigue develops?

_____

_____

2. What would you recommend to an interested friend as the best ways to delay the onset of fatigue?

_____

_____

_____

The Skeletal Muscle Length-Tension Relationship

OBJECTIVES

1. To understand the terms *isometric contraction, active force, passive force, total force,* and *length-tension relationship.*

2. To understand the effect that resting muscle length has on tension development when the muscle is maximally stimulated in an isometric experiment.

3. To explain the molecular basis of the skeletal muscle length-tension relationship.

Introduction

Skeletal muscle contractions are either isometric or isotonic. When a muscle attempts to move a load that is equal to the force generated by the muscle, the muscle contracts isometrically. During an **isometric** contraction, the muscle stays at a fixed length (*isometric* means "same length"). An example of isometric muscle contraction is when you stand in a doorway and push on the doorframe. The load that you are attempting to move (the doorframe) can easily equal the force generated by your muscles, so your muscles do not shorten even though they are actively contracting.

Isometric contractions are accomplished experimentally by keeping both ends of the muscle in a fixed position while electrically stimulating the muscle. Resting length (the length of the muscle before stimulation) is an important factor in determining the amount of force that a muscle can develop when stimulated. **Passive force** is generated by stretching the muscle and results from the elastic recoil of the tissue itself. This passive force is largely caused by the protein titin, which acts as a molecular bungee cord. **Active force** is generated when myosin thick filaments bind to actin thin filaments, thus engaging the cross bridge cycle and ATP hydrolysis. Think of the skeletal muscle as having two force properties: it exerts passive force when it is stretched (like a rubber band exerts passive force) and active force when it is stimulated. **Total force** is the sum of passive and active forces.

This activity allows you to set and hold constant the length of the isolated skeletal muscle and subsequently stimulate it with individual maximal voltage stimuli. A graph relating the three forces generated and the fixed length of the muscle will be automatically plotted after you stimulate the muscle. In muscle physiology this graph is known as the **isometric length-tension relationship.** The results of this simulation can be applied to human muscles to understand how optimum resting length will result in maximum force production.

To understand why muscle tissue behaves as it does, you must understand tension at the cellular level. If you have difficulty understanding the results of this activity, review the sliding filament model of muscle contraction. Think of the length-tension relationship in terms of those sarcomeres that are too short, those that are too long, and those that have the ideal amount of thick and thin filament overlap.

> **EQUIPMENT USED** The following equipment will be depicted on-screen: intact, viable skeletal muscle dissected off the leg of a frog; electrical stimulator—delivers the desired amount and duration of stimulating voltage to the muscle via electrodes resting on the muscle; mounting stand—includes (1) a force transducer to measure the amount of force, or tension, developed by the muscle and (2) a gearing system that allows the hook through the muscle's lower tendon to be moved up or down, thus altering the fixed length of the muscle; oscilloscope—displays the stimulated muscle twitch and the amount of active, passive, and total force developed by the muscle.

Experiment Instructions

Go to the home page in the PhysioEx software and click **Exercise 2, Skeletal Muscle Physiology.** Click **Activity 6, The Skeletal Muscle Length-Tension Relationship,** and take the online **Pre-lab Quiz** for Activity 6.

After you take the online Pre-lab Quiz, click the **Experiment** tab and begin the experiment. The experiment instructions are reprinted here for your reference. The opening screen for the experiment is shown below.

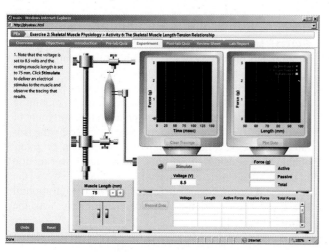

1. Note that the voltage is set to 8.5 volts and the resting muscle length is set to 75 mm. Click **Stimulate** to deliver an electrical stimulus to the muscle and observe the tracing that results.

2. You should see a single muscle twitch tracing on the left oscilloscope display and three data points (representing active, passive, and total force generated during this twitch) plotted on the right display. The yellow box represents the total force, the red dot contained within the yellow box represents the active force, and the green square represents the passive force. Click **Record Data** to display your results in the grid (and record your results in Chart 6).

CHART 6	Skeletal Muscle Length-Tension Relationship		
Length (mm)	Active force (g)	Passive force (g)	Total force (g)

> **? PREDICT Question 1**
> As the resting length of the muscle is changed, what will happen to the amount of total force the muscle generates during the stimulated twitch?
> _____
> _____

3. You will now gradually shorten the muscle to determine the effect of muscle length on active, passive, and total force.

- Shorten the muscle by 5 mm by clicking the − button beside the muscle length display.

- Click **Stimulate** to deliver an electrical stimulus to the muscle and note the values of the total, active, and passive forces relative to those observed at the original 75 mm.

- Click **Record Data** to display your results in the grid (and record your results in Chart 6).

Repeat these steps until you reach a muscle length of 50 mm.

4. Click **Clear Tracings** to clear the left oscilloscope display.

5. Lengthen the muscle to 80 mm by clicking the + button beside the muscle length display. Click **Stimulate** to deliver an electrical stimulus to the muscle and note the values of the total, active, and passive forces relative to those observed at the original 75 mm.

6. Click **Record Data** to display your results in the grid (and record your results in Chart 6).

7. You will now gradually lengthen the muscle to determine the effect of muscle length on active, passive, and total force.

- Lengthen the muscle by 10 mm by clicking the + button beside the muscle length display.

- Click **Stimulate** to deliver an electrical stimulus to the muscle and note the values of the total, active, and passive forces relative to those observed at the original 75 mm.

- Click **Record Data** to display your results in the grid (and record your results in Chart 6).

Repeat these steps until you reach a muscle length of 100 mm.

8. Click **Plot Data** to view a summary of your data on a plotted grid. Click **Submit** to record your plot in the lab report.

After you complete the experiment, take the online **Post-lab Quiz** for Activity 6.

Activity Questions

1. Explain what happens in the skeletal muscle sarcomere to result in the changes in active, passive, and total force when the resting muscle length is changed.

_____

_____

2. Explain the dip in the total force curve as the muscle was stretched to longer lengths. (Hint: Keep in mind that you are measuring the sum of active and passive forces.)

_____

_____

Isotonic Contractions and the Load-Velocity Relationship

OBJECTIVES

1. To understand the terms *isotonic concentric contraction, load, latent period, shortening velocity,* and *load-velocity relationship.*

2. To understand the effect that increasing load (that is, weight) has on an isolated skeletal muscle when the muscle is stimulated in an isotonic contraction experiment.

3. To understand the load-velocity relationship in isolated skeletal muscle.

Introduction

Skeletal muscle contractions can be described as either isometric or isotonic. When a muscle attempts to move an object (the **load**) that is equal in weight to the force generated by the muscle, the muscle is observed to contract isometrically. In an isometric contraction, the muscle stays at a fixed length (*isometric* means "same length").

During an **isotonic contraction,** the skeletal muscle length changes and, thus, the load moves a measurable distance. If the muscle length shortens as the load moves, the contraction is called an **isotonic *concentric* contraction.** An isotonic concentric contraction occurs when a muscle generates a force greater than the load attached to the muscle's end. In this type of contraction, there is a **latent period** during which there is a rise in muscle tension but no observable movement of the weight. After the muscle tension exceeds the weight of the load, an isotonic concentric contraction can begin. Thus, the latent period gets longer as the weight of the load gets larger. When the building muscle force exceeds the load, the muscle shortens and the weight moves. Eventually, the force of the muscle contraction will decrease as the muscle twitch begins the relaxation phase, and the load will therefore start to return to its original position.

An isotonic twitch is not an all-or-nothing event. If the load is increased, the muscle must generate more force to move it and the latent period will therefore get longer because it will take more time for the necessary force to be generated by the muscle. The speed of the contraction (muscle **shortening velocity**) also depends on the load that the muscle is attempting to move. Maximal shortening velocity is attained with minimal load attached to the muscle. Conversely, the heavier the load, the slower the muscle twitch. You can think of lifting an object from the floor as an example. A light object can be lifted quickly (high velocity), whereas a heavier object will be lifted with a slower velocity for a shorter duration.

In an isotonic muscle contraction experiment, one end of the muscle remains free (unlike in an isometric contraction experiment, where both ends of the muscle are held in a fixed position). Different weights (loads) can then be attached to the free end of the isolated muscle, while the other end is held in a fixed position by the force transducer. If the weight (the load) is less than the tension generated by the whole muscle, then the muscle will be able to lift it with a measurable distance, velocity, and duration. In this activity, you will change the weight (load) that the muscle will try to move as it shortens.

> **EQUIPMENT USED** The following equipment will be depicted on-screen: intact, viable skeletal muscle dissected off the leg of a frog; electrical stimulator—delivers the desired amount and duration of stimulating voltage to the muscle via electrodes resting on the muscle; mounting stand—includes a ruler that allows a rapid measurement of the distance (cm) that the weight (load) is lifted by the isolated muscle; several weights (in grams)—can be interchangeably attached to the hook on the free lower tendon of the mounted skeletal muscle; oscilloscope—displays the stimulated isotonic concentric contraction, the duration of the contraction, and the distance that muscle lifts the weight (load).

Experiment Instructions

Go to the home page in the PhysioEx software and click **Exercise 2: Skeletal Muscle Physiology.** Click **Activity 7: Isotonic Contractions and the Load-Velocity Relationship,** and take the online **Pre-lab Quiz** for Activity 7.

  After you take the online Pre-lab Quiz, click the **Experiment** tab and begin the experiment. The experiment instructions are reprinted here for your reference. The opening screen for the experiment is shown below.

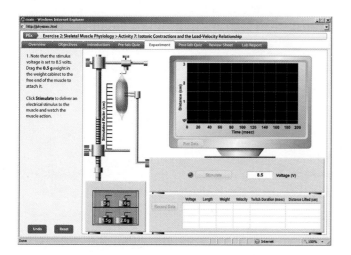

1. Note that the stimulus voltage is set to 8.5 volts. Drag the 0.5-g weight in the weight cabinet to the free end of the muscle to attach it. Click **Stimulate** to deliver an electrical stimulus to the muscle and watch the muscle action.

2. Observe that, as the muscle shortens in length, it lifts the weight off the platform. The muscle then lengthens as it relaxes and lowers the weight back down to the platform. Click **Stimulate** again and try to watch both the muscle and the oscilloscope screen at the same time.

3. Click **Record Data** to display your results in the grid (and record your results in Chart 7).

CHART 7	Isotonic Contraction Results		
Weight (g)	Velocity (cm/sec)	Twitch duration (msec)	Distance lifted (cm)

PREDICT Question 1
As the load on the muscle *increases*, what will happen to the latent period, the shortening velocity, the distance that the weight moved, and the contraction duration?

_____

_____

4. Remove the 0.5-g weight by dragging it back to the weight cabinet. Drag the 1.0-g weight to the free end of the muscle to attach it. Click **Stimulate** and observe the muscle and the oscilloscope screen.

5. Click **Record Data** to display your results in the grid (and record your results in Chart 7).

6. Remove the 1.0-g weight by dragging it back to the weight cabinet. Drag the 1.5-g weight to the free end of the muscle to attach it. Click **Stimulate** and observe the muscle and the oscilloscope screen.

7. Click **Record Data** to display your results in the grid (and record your results in Chart 7).

8. Remove the 1.5-g weight by dragging it back to the weight cabinet. Drag the 2.0-g weight to the free end of the muscle to attach it. Click **Stimulate** and observe the muscle and the oscilloscope screen.

9. Click **Record Data** to display your results in the grid (and record your results in Chart 7).

10. Click **Plot Data** to generate a muscle load-velocity relationship. Watch the display carefully as the program animates the development of a load-velocity relationship for the data you have collected. Click **Submit** to record your plot in the lab report.

After you complete the experiment, take the online **Post-lab Quiz** for Activity 7.

Activity Questions

1. Explain the relationship between the load attached to a skeletal muscle and the initial velocity of skeletal muscle shortening.

_____

_____

_____

2. Explain why it will take you longer to perform ten repetitions lifting a 20-pound weight than it would to perform the same number of repetitions with a 5-pound weight.

_____

_____

_____

NAME_____

LAB TIME/DATE _____

Skeletal Muscle Physiology

ACTIVITY 1 The Muscle Twitch and the Latent Period

1. Define the terms *skeletal muscle fiber, motor unit, skeletal muscle twitch, electrical stimulus,* and *latent period.* _____

_____

_____

_____

_____

2. What is the role of acetylcholine in a skeletal muscle contraction? _____

_____

3. Describe the process of excitation-contraction coupling in skeletal muscle fibers. _____

_____

_____

4. Describe the three phases of a skeletal muscle twitch. _____

_____

_____

_____

5. Does the duration of the latent period change with different stimulus voltages? How well did the results compare with your

   prediction? _____

_____

6. At the threshold stimulus, do sodium ions start to move into or out of the cell to bring about the membrane depolarization?

_____

_____

ACTIVITY 2 The Effect of Stimulus Voltage on Skeletal Muscle Contraction

1. Describe the effect of increasing stimulus voltage on isolated skeletal muscle. Specifically, what happened to the muscle force generated with stronger electrical stimulations and why did this change occur? How well did the results compare with your prediction? _____

   _____

2. How is this change in whole-muscle force achieved in vivo? _____

   _____

3. What happened in the isolated skeletal muscle when the maximal voltage was applied? _____

   _____

ACTIVITY 3 The Effect of Stimulus Frequency on Skeletal Muscle Contraction

1. What is the difference between stimulus intensity and stimulus frequency? _____

2. In this experiment you observed the effect of stimulating the isolated skeletal muscle multiple times in a short period with complete relaxation between the stimuli. Describe the force of contraction with each subsequent stimulus. Are these results called treppe or wave summation? _____

   _____

3. How did the frequency of stimulation affect the amount of force generated by the isolated skeletal muscle when the frequency of stimulation was increased such that the muscle twitches did not fully relax between subsequent stimuli? Are these results called treppe or wave summation? How well did the results compare with your prediction? _____

   _____

   _____

   _____

4. To achieve an active force of 5.2 g, did you have to increase the stimulus voltage above 8.5 volts? If not, how did you achieve an active force of 5.2 g? How well did the results compare with your prediction? _____

   _____

   _____

5. Compare and contrast frequency-dependent wave summation with motor unit recruitment (previously observed by increasing the stimulus voltage). How are they similar? How was each achieved in the experiment? Explain how each is achieved in vivo. _____

   _____

ACTIVITY 4   Tetanus in Isolated Skeletal Muscle

1. Describe how increasing the stimulus frequency affected the force developed by the isolated whole skeletal muscle in this activity. How well did the results compare with your prediction? _____

   _____

2. Indicate what type of force was developed by the isolated skeletal muscle in this activity at the following stimulus frequencies: at 50 stimuli/sec, at 140 stimuli/sec, and above 146 stimuli/sec. _____

   _____

3. Beyond what stimulus frequency is there no further increase in the peak force? What is the muscle tension called at this frequency? _____

ACTIVITY 5   Fatigue in Isolated Skeletal Muscle

1. When a skeletal muscle fatigues, what happens to the contractile force over time? _____

   _____

2. What are some proposed causes of skeletal muscle fatigue? _____

   _____

3. Turning the stimulator off allows a small measure of muscle recovery. Thus, the muscle will produce more force for a longer time period if the stimulator is briefly turned off than if the stimuli were allowed to continue without interruption. Explain why this might occur. How well did the results compare with your prediction? _____

   _____

   _____

   _____

4. List a few ways that humans could delay the onset of fatigue when they are vigorously using their skeletal muscles. _____

   _____

ACTIVITY 6   The Skeletal Muscle Length-Tension Relationship

1. What happens to the amount of total force the muscle generates during the stimulated twitch? How well did the results compare with your prediction? _____

   _____

   _____

2. What is the key variable in an isometric contraction of a skeletal muscle? _____

3. Based on the unique arrangement of myosin and actin in skeletal muscle sarcomeres, explain why active force varies with

   changes in the muscle's resting length. _____

   _____

   _____

4. What skeletal muscle lengths generated passive force? (Provide a range.) _____

5. If you were curling a 7-kg dumbbell, when would your bicep muscles be contracting isometrically? _____

ACTIVITY 7   Isotonic Contractions and the Load-Velocity Relationship

1. If you were using your bicep muscles to curl a 7-kg dumbbell, when would your muscles be contracting isotonically?

   _____

2. Explain why the latent period became longer as the load became heavier in the experiment. How well did the results com-

   pare with your prediction? _____

   _____

   _____

3. Explain why the shortening velocity became slower as the load became heavier in this experiment. How well did the results

   compare with your prediction? _____

   _____

   _____

4. Describe how the shortening distance changed as the load became heavier in this experiment. How well did the results com-

   pare with your prediction? _____

   _____

   _____

5. Explain why it would take you longer to perform 10 repetitions lifting a 10-kg weight than it would to perform the same num-

   ber of repetitions with a 5-kg weight. _____

   _____

6. Describe what would happen in the following experiment: A 2.5-g weight is attached to the end of the isolated whole skele-

   tal muscle used in these experiments. Simultaneously, the muscle is maximally stimulated by 8.5 volts and the platform sup-

   porting the weight is removed. Will the muscle generate force? Will the muscle change length? What is the name for this type

   of contraction? _____

   _____

Neurophysiology of Nerve Impulses

Exercise Overview

The nervous system contains two general types of cells: **neurons** and neuroglia (or glial cells). This exercise focuses on neurons. Neurons respond to their local environment by generating an electrical signal. For example, sensory neurons in the nose generate a signal (called a **receptor potential**) when odor molecules interact with receptor proteins on the membrane of these olfactory sensory neurons. Thus, sensory neurons can respond directly to sensory stimuli. The receptor potential can trigger another electrical signal (called an **action potential**), which travels along the membrane of the sensory neuron's axon to the brain—you could say that the action potential is conducted to the brain.

The action potential causes the release of **chemical neurotransmitters** onto neurons in olfactory regions of the brain. These chemical neurotransmitters bind to receptor proteins on the membrane of these brain **interneurons.** In general, interneurons respond to chemical neurotransmitters released by other neurons. In the nose the odor molecules are sensed by sensory neurons. In the brain the odor is perceived by the activity of interneurons responding to neurotransmitters. Any resulting action or behavior is caused by the subsequent activity of **motor neurons,** which can stimulate muscles to contract (see Exercise 2).

In general each neuron has three functional regions for signal transmission: a receiving region, a conducting region, and an output region, or secretory region. Sensory neurons often have a receptive ending specialized to detect a specific sensory stimulus, such as odor, light, sound, or touch. The **cell body** and **dendrites** of interneurons receive stimulation by neurotransmitters at structures called **chemical synapses** and produce **synaptic potentials.** The conducting region is

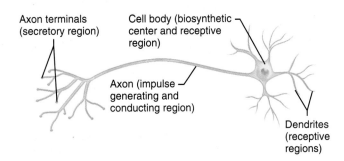

Axon terminals (secretory region)

Cell body (biosynthetic center and receptive region)

Axon (impulse generating and conducting region)

Dendrites (receptive regions)

FIGURE 3.1   A neuron with functional areas identified.

usually an **axon,** which ends in an output region (the axon terminal) where neurotransmitter is released (view Figure 3.1).

Although the neuron is a single cell surrounded by a continuous plasma membrane, each region contains distinct membrane proteins that provide the basis for the functional differences. Thus, the receiving end has receptor proteins and proteins that generate the receptor potential, the conducting region has proteins that generate and conduct action potentials, and the output region has proteins to package and release neurotransmitters. Membrane proteins are found throughout the neuronal membrane—many of these proteins transport ions (see Exercise 1).

The signals generated and conducted by neurons are electrical. In ordinary household devices, electric current is carried by electrons. In biological systems, currents are carried by positively or negatively charged **ions.** Like charges repel each other and opposite charges attract. In general, ions cannot easily pass through the lipid bilayer of the plasma membrane and must pass through **ion channels** formed by integral membrane proteins. Some channels are usually open (leak channels) and others are gated, meaning that the channel can be in an open or closed configuration. Channels can also be selective for which ions are allowed to pass. For example, sodium channels are mostly permeable to sodium ions when open, and potassium channels are mostly permeable to potassium ions when open. The term **conductance** is often used to describe **permeability.** In general, ions will flow through an open channel from a region of higher concentration to a region of lower concentration (see Exercise 1). In this exercise you will explore some of these characteristics applied to neurons.

Although it is possible to measure the ionic currents through the membrane (even the currents passing through single ion channels), it is more common to measure the potential difference, or voltage, across the membrane. This membrane voltage is usually called the **membrane potential,** and the units are **millivolts (mV).** One can think of the membrane as a battery, a device that separates and stores charge. A typical household battery has a positive and a negative pole so that when it is connected, for example through a lightbulb in a flashlight, current flows through the bulb. Similarly, the plasma membrane can store charge and has a relatively positive side and a relatively negative side. Thus, the membrane is said to be **polarized.** When these two sides (intracellular and extracellular) are connected through open ion channels, current in the form of ions can flow in or out across the membrane and thus change the membrane voltage.

The Resting Membrane Potential

OBJECTIVES

1. To define the term *resting membrane potential.*
2. To measure the resting membrane potential in different parts of a neuron.
3. To determine how the resting membrane potential depends on the concentrations of potassium and sodium.
4. To understand the ion conductances/ion channels involved in the resting membrane potential.

Introduction

The receptor potential, synaptic potentials, and action potentials are important signals in the nervous system. These potentials refer to changes in the membrane potential from its resting level. In this activity you will explore the nature of the resting potential. The **resting membrane potential** is really a potential difference between the inside of the cell (intracellular) and the outside of the cell (extracellular) across the membrane. It is a steady-state condition that depends on the resting permeability of the membrane to ions and on the intracellular and extracellular concentrations of those ions to which the membrane is permeable.

For many neurons, Na^+ and K^+ are the most important ions, and the concentrations of these ions are established by transport proteins, such as the Na^+-K^+ pump, so that the intracellular Na^+ concentration is low and the intracellular K^+ concentration is high. Inside a typical cell, the concentration of K^+ is ~150 mM and the concentration of Na^+ is ~5 mM. Outside a typical cell, the concentration of K^+ is ~5 mM and the concentration of Na^+ is ~150 mM. If the membrane is permeable to a particular ion, that ion will diffuse down its concentration gradient from a region of higher concentration to a region of lower concentration. In the generation of the resting membrane potential, K^+ ions diffuse out across the membrane, leaving behind a net negative charge—large anions that cannot cross the membrane.

The membrane potential can be measured with an amplifier. In the experiment the extracellular solution is connected to a ground (literally, the earth) which is defined as 0 mV. To record the voltage across the membrane, a microelectrode is inserted through the membrane without significantly damaging it. Typically, the microelectrode is made by pulling a thin glass pipette to a fine hollow point and filling the pulled pipette with a salt solution. The salt solution conducts electricity like a wire, and the glass insulates it. Only the tip of the microelectrode is inserted through the membrane, and the filled tip of the microelectrode makes electrical contact with the intracellular solution. A wire connects the microelectrode to the input of the amplifier so that the amplifier records the membrane potential, the voltage across the membrane between the intracellular and grounded extracellular solutions.

The membrane potential and the various signals can be observed on an oscilloscope. An electron beam is pulled up or down according to the voltage as it sweeps across a phosphorescent screen. Voltages below 0 mV are negative and voltages above 0 mV are positive. For this first activity, the time of the sweep is set for 1 second per division, and the sensitivity is set to 10 mV per division; a division is the distance between gridlines on the oscilloscope.

EQUIPMENT USED The following equipment will be depicted on-screen: neuron (in vitro)—a large, dissociated (or cultured) neuron; three extracellular solutions—control, high potassium, and low sodium; microelectrode—a probe with a very small tip that can impale a single neuron (In an actual wet lab, a microelectrode manipulator is used to position the microelectrode. For simplicity, the microelectrode manipulator will not be depicted in this activity.); microelectrode manipulator controller—controls movement of the manipulator; microelectrode amplifier—used to measure the voltage between the microelectrode and a reference; oscilloscope—used to observe voltage changes.

Experiment Instructions

Go to the home page in the PhysioEx software, and click **Exercise 3: Neurophysiology of Nerve Impulses.** Click **Activity 1: The Resting Membrane Potential,** and take the online **Pre-lab Quiz** for Activity 1.

After you take the online Pre-lab Quiz, click the **Experiment** tab and begin the experiment. The experiment instructions are reprinted here for your reference. The opening screen for the experiment is shown below.

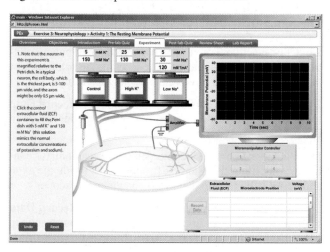

1. Note that the neuron in this experiment is magnified relative to the petri dish. In a typical neuron, the cell body, which is the thickest part, is 5–100 μm wide, and the axon might be only 0.5 μm wide.

Click the *control* **extracellular fluid (ECF)** container to fill the petri dish with 5 mM K$^+$ and 150 mM Na$^+$ (this solution mimics the normal extracellular concentrations of potassium and sodium).

2. Note that a reference electrode is already positioned in the petri dish. This reference electrode is connected to ground through the amplifier.

Click position **1** on the microelectrode manipulator controller to position the microelectrode tip in the solution, just outside the cell body, and observe the tracing that results on the oscilloscope.

3. Note the oscilloscope tracing of the voltage outside the cell body and click **Record Data** to display your results in the grid (and record your results in Chart 1).

4. Click position **2** on the microelectrode manipulator controller to position the microelectrode tip just inside the cell body and observe the tracing that results.

5. Note the oscilloscope tracing of the voltage inside the cell body and click **Record Data** to display your results in the grid (and record your results in Chart 1). This is the resting membrane potential; that is, the potential difference between intracellular and extracellular membrane voltages. By convention, the extracellular resting membrane voltage is taken to be 0 mV.

6. Click position **3** on the microelectrode manipulator controller to position the microelectrode tip in the solution, just outside the axon, and observe the tracing that results.

7. Note the oscilloscope tracing of the voltage outside the axon and click **Record Data** to display your results in the grid (and record your results in Chart 1).

CHART 1	Resting Membrane Potential	
Extracellular fluid (ECF)	**Microelectrode position**	**Voltage (mV)**

8. Click position **4** on the microelectrode manipulator controller to position the microelectrode tip just inside the axon and observe the tracing that results.

9. Note the oscilloscope tracing of the voltage inside the axon and click **Record Data** to display your results in the grid (and record your results in Chart 1).

? PREDICT Question 1
Predict what will happen to the resting membrane potential if the extracellular K^+ concentration is increased.

_____

_____

10. You will now change the concentrations of the ions in the extracellular fluid to determine which ions contribute most to the separation of charge across the membrane. The extracellular potassium concentration is normally low, so you will first increase the extracellular potassium concentration.

In the high K^+ ECF solution the K^+ concentration has been increased fivefold, from 5 to 25 mM. To keep the number of positive charges in the extracellular solution constant, the Na^+ concentration has been reduced by 20 mM, from 150 to 130 mM. As you will see, this relatively small decrease in Na^+ will not by itself change the membrane potential. Note that in this activity, the generation of the action potential (which is covered in Activities 3–9) is blocked with a toxin. Click the **high K^+ ECF** container to change the solution in the petri dish to 25 mM K^+ and 130 mM Na^+.

11. Note the voltage inside the axon and click **Record Data** to display your results in the grid (and record your results in Chart 1).

12. Click position **3** on the microelectrode manipulator controller to position the microelectrode tip in the solution, just outside the axon, and observe the tracing that results.

13. Note the voltage outside the axon and click **Record Data** to display your results in the grid (and record your results in Chart 1).

14. Click position **1** on the microelectrode manipulator controller to position the microelectrode tip in the solution, just outside the cell body, and observe the tracing that results.

15. Note the voltage outside the cell body and click **Record Data** to display your results in the grid (and record your results in Chart 1).

16. Click position **2** on the microelectrode manipulator controller to position the microelectrode tip just inside the cell body and observe the tracing that results on the oscilloscope.

17. Note the voltage inside the cell body and click **Record Data** to display your results in the grid (and record your results in Chart 1).

18. Click the *control* ECF container to change back to the normal K^+ concentration and note the change in voltage inside the cell body.

19. You will now decrease the extracellular Na^+ concentration (the extracellular Na^+ concentration is normally high).

The extracellular sodium concentration in the low Na^+ solution has been decreased fivefold, from 150 mM to 30 mM. To keep the number of positive charges constant in the extracellular solution, the Na^+ has been replaced by the same amount of a large monovalent cation. Note that the extracellular Na^+ concentration, even in the low Na^+ ECF, is higher than the intracellular Na^+ concentration. Click the **low Na^+ ECF** container to change the solution in the petri dish to 5 mM K^+ and 30 mM Na^+.

20. Note the voltage inside the cell body and click **Record Data** to display your results in the grid (and record your results in Chart 1).

21. Click position **1** on the microelectrode manipulator controller to position the microelectrode tip in the solution, just outside the cell body, and observe the tracing that results.

22. Note the voltage outside the cell body and click **Record Data** to display your results in the grid (and record your results in Chart 1).

23. Click position **3** on the microelectrode manipulator controller to position the microelectrode tip in the solution, just outside the axon, and observe the tracing that results.

24. Note the voltage outside the axon and click **Record Data** to display your results in the grid (and record your results in Chart 1).

25. Click position **4** on the microelectrode manipulator controller to position the microelectrode tip just inside the axon and observe the tracing that results on the oscilloscope.

26. Note the voltage inside the axon and click **Record Data** to display your results in the grid (and record your results in Chart 1).

After you complete the experiment, take the online **Post-lab Quiz** for Activity 1.

Activity Questions

1. Explain why the resting membrane potential had the same value in the cell body and in the axon.

_____

_____

2. Describe what would happen to a resting membrane potential if the sodium-potassium transport pump was blocked.

_____

_____

3. Describe what would happen to a resting membrane potential if the concentration of large intracellular anions that are unable to cross the membrane is experimentally increased.

Receptor Potential

OBJECTIVES

1. To define the terms *sensory receptor, receptor potential, sensory transduction, stimulus modality,* and *depolarization.*
2. To determine the *adequate stimulus* for different sensory receptors.
3. To demonstrate that the receptor potential amplitude increases with stimulus intensity.

Introduction

The receiving end of a sensory neuron, the **sensory receptor,** has receptor proteins (as well as other membrane proteins) that can generate a signal called the **receptor potential** when the sensory neuron is stimulated by an appropriate, adequate stimulus. In this activity you will use the same recording instruments and microelectrode that you used in Activity 1. However, in this activity, you will record from the sensory receptor of three different sensory neurons and examine how these neurons respond to sensory stimuli of different modalities.

The sensory region will be shown disconnected from the rest of the neuron so that you can record the receptor potential in isolation. Similar results can sometimes be obtained by treating a whole neuron with chemicals that block the responses generated by the axon. The molecules localized to the sensory receptor ending are able to generate a receptor potential when an adequate stimulus is applied. The energy in the stimulus (for example, chemical, physical, or heat) is changed into an electrical response that involves the opening or closing of membrane ion channels. The general process that produces this change is called **sensory transduction,** which occurs at the receptor ending of the sensory neuron. Sensory transduction can be thought of as a type of signal transduction where the signal is the sensory stimulus.

You will observe that, with an appropriate stimulus, the amplitude of the receptor potential increases with stimulus intensity. Such a response is an example of a potential that is graded with stimulus intensity. These responses are sometimes referred to as *graded potentials,* or *local potentials.* Thus, the receptor potential is a graded, or local, potential. If the response (receptor potential) is a change in membrane potential from the negative resting potential to a less negative level, the membrane becomes less polarized and the change is called **depolarization.**

EQUIPMENT USED The following equipment will be depicted on-screen: three sensory receptors—Pacinian (lamellar) corpuscle, olfactory receptor, and free nerve ending; microelectrode—a probe with a very small tip that can impale a single neuron (In an actual wet lab, a microelectrode manipulator is used to position the microelectrodes. For simplicity, the microelectrode manipulator will not be depicted in this activity.); microelectrode amplifier—used to measure the voltage between the microelectrode and a reference; stimulator—used to select the stimulus modality (pressure, chemical, heat, or light) and intensity (low, moderate, or high); oscilloscope—used to observe voltage changes.

Experiment Instructions

Go to the home page in the PhysioEx software, and click **Exercise 3: Neurophysiology of Nerve Impulses.** Click **Activity 2: Receptor Potential,** and take the online **Pre-lab Quiz** for Activity 2.

After you take the online Pre-lab Quiz, click the **Experiment** tab and begin the experiment. The experiment instructions are reprinted here for your reference. The opening screen for the experiment is shown below.

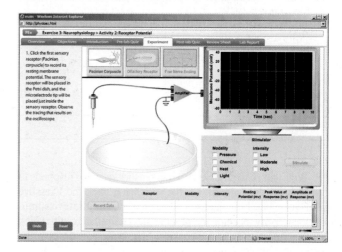

1. Note that the timescale on the oscilloscope has been changed from 1 second per division to 10 milliseconds per division, so that you can observe the responses recorded in the sensory receptors more clearly. Click the first sensory receptor (Pacinian corpuscle) to record its resting membrane potential. The sensory receptor will be placed in the petri dish, and the microelectrode tip will be placed just inside the sensory receptor. Observe the tracing that results on the oscilloscope.

2. Note the voltage inside the sensory receptor and click **Record Data** to display your results in the grid (and record your results in Chart 2).

? PREDICT Question 1
The adequate stimulus for a Pacinian corpuscle is pressure or vibration on the skin. For a Pacinian corpuscle, which modality will induce a receptor potential of the largest amplitude?

CHART 2	Receptor Potential		
		Receptor potential (mV)	
Stimulus modality	Pacinian (lamellar) corpuscle	Olfactory receptor	Free nerve ending
None			
Pressure			
Low			
Moderate			
High			
Chemical			
Low			
Moderate			
High			
Heat			
Low			
Moderate			
High			
Light			
Low			
Moderate			
High			

3. You will now observe how the sensory receptor responds to different sensory stimuli. On the stimulator, click the **Pressure** modality. Click **Low** intensity and then click **Stimulate** to stimulate the sensory receptor and observe the tracing that results. Click **Moderate** intensity and then click **Stimulate** and observe the tracing that results. Click **High** intensity and then click **Stimulate** and observe the tracing that results. Click **Record Data** to display your results in the grid (and record your results in Chart 2).

4. On the stimulator, click the **Chemical** (odor) modality. Click **Low** intensity and then click **Stimulate** to stimulate the sensory receptor and observe the tracing that results. Click **Moderate** intensity and then click **Stimulate** and observe the tracing that results. Click **High** intensity and then click **Stimulate** and observe the tracing that results. Click **Record Data** to display your results in the grid (and record your results in Chart 2).

5. On the stimulator, click the **Heat** modality. Click **Low** intensity and then click **Stimulate** to stimulate the sensory receptor and observe the tracing that results. Click **Moderate** intensity and then click **Stimulate** and observe the tracing that results. Click **High** intensity and then click **Stimulate** and observe the tracing that results. Click **Record Data** to display your results in the grid (and record your results in Chart 2).

6. On the stimulator, click the **Light** modality. Click **Low** intensity and then click **Stimulate** to stimulate the sensory receptor and observe the tracing that results. Click **Moderate** intensity and then click **Stimulate** and observe the tracing that results. Click **High** intensity and then click **Stimulate** and observe the tracing that results. Click **Record Data** to display your results in the grid (and record your results in Chart 2).

? PREDICT Question 2
The adequate stimuli for olfactory receptors are chemicals, typically odorant molecules. For an olfactory receptor, which modality will induce a receptor potential of the largest amplitude?

_____

7–12. Repeat steps 1–6 with the next sensory receptor: olfactory receptor.

13–18. Repeat steps 1–6 with the next sensory receptor: free nerve ending.

After you complete the experiment, take the online **Post-lab Quiz** for Activity 2.

Activity Questions

1. Are graded receptor potentials always depolarizing? Do graded receptor potentials always make it easier to induce action potentials?

_____

_____

2. Based on the definition of membrane depolarization in this activity, define membrane *hyperpolarization*.

_____

_____

3. What do you think is the adequate stimulus for sensory receptors in the ear? Can you think of a stimulus that would inappropriately activate the sensory receptors in the ear if the stimulus had enough intensity?

_____

_____ ▬

ACTIVITY 3

The Action Potential: Threshold

OBJECTIVES

1. To define the terms *action potential, nerve, axon hillock, trigger zone,* and *threshold*.
2. To predict how an increase in extracellular K^+ could trigger an action potential.

Introduction

In this activity you will explore changes in potential that occur in the axon. Axons are long, thin structures that conduct a signal called the **action potential.** A **nerve** is a bundle of axons.

Axons are typically studied in a nerve chamber. In this activity the axon will be draped over wires that make electrical contact with the axon and can therefore record the electrical activity in the axon. Because the axon is so thin, it is very difficult to insert an electrode across the membrane into the axon. However, some of the charge (ions) that crosses the membrane to generate the action potential can be recorded from outside the membrane (extracellular recording), as you will do in this activity. The molecular mechanisms underlying the action potential were explored more than 50 years ago with intracellular recording using the giant axons of the squid, which are about 1 millimeter in diameter.

In this activity the axon will be artificially disconnected from the cell body and dendrites. In a typical multipolar neuron (view Figure 3.1 in the Exercise Overview), the axon extends from the cell body at a region called the **axon hillock.** In a myelinated axon, this first region is called the initial segment. An action potential is usually initiated at the junction of the axon hillock and the initial segment; therefore, this region is also referred to as the **trigger zone.**

You will use an electrical stimulator to explore the properties of the action potential. Current passes from the stimulator to one of the stimulation wires, then across the axon, and then back to the stimulator through a second wire. This current will depolarize the axon. Normally, in a sensory neuron, the depolarizing receptor potential spreads passively to the axon hillock and produces the depolarization needed to evoke the action potential. Once an action potential is generated, it is regenerated down the membrane of the axon. In other words, the action potential is **propagated,** or *conducted,* down the axon (see Activity 6).

You will now generate an action potential at one end of the axon by stimulating it electrically and record the action potential that is propagated down the axon. The extracellular action potential that you record is similar to one that would be recorded across the membrane with an intracellular microelectrode, but much smaller. For simplicity, only one axon is depicted in this activity.

> **EQUIPMENT USED** The following equipment will be depicted on-screen: nerve chamber; axon; oscilloscope—used to observe timing of stimuli and voltage changes in the axon; stimulator—used to set the stimulus voltage and to deliver pulses that depolarize the axon; stimulation wires (S); recording electrodes (wires R1 and R2)—used to record voltage changes in the axon. (The first set of recording electrodes, R1, is 2 centimeters from the stimulation wires, and the second set of recording electrodes, R2, is 2 centimeters from R1.)

Experiment Instructions

Go to the home page in the PhysioEx software, and click **Exercise 3: Neurophysiology of Nerve Impulses.** Click **Activity 3: The Action Potential: Threshold,** and take the online **Pre-lab Quiz** for Activity 3.

After you take the online Pre-lab Quiz, click the **Experiment** tab and begin the experiment. The experiment instructions are reprinted here for your reference. The opening screen for the experiment is shown below.

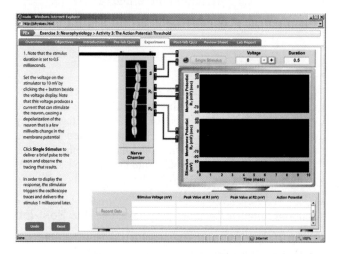

1. Note that the stimulus duration is set to 0.5 milliseconds. Set the voltage on the stimulator to 10 mV by clicking the + button beside the voltage display. Note that this voltage produces a current that can stimulate the neuron, causing a depolarization of the neuron that is a change of a few millivolts in the membrane potential.

Click **Single Stimulus** to deliver a brief pulse to the axon and observe the tracing that results. In order to display the response, the stimulator triggers the oscilloscope traces and delivers the stimulus 1 millisecond later.

2. Note that the recording electrodes R1 and R2 record the extracellular voltage, rather than the actual membrane potential. The 10 mV depolarization at the site of stimulation only occurs locally at that site and is not recorded farther down the axon. At this initial stimulus voltage, there was no action potential. Click **Record Data** to display your results in the grid (and record your results in Chart 3).

CHART 3	Threshold			
Stimulus voltage (mV)	Peak value at R1 (µV)	Peak value at R2 (µV)	Action potential	

3. You will increase the stimulus voltage until you observe an action potential at recording electrode 1 (R1). Increase the voltage by 10 mV by clicking the **+** button beside the voltage display and then click **Single Stimulus.** The voltage at which you first observe an action potential is the **threshold voltage.** Note that the action potential recorded extracellularly is quite small. Intracellularly, the membrane potential would change from −70 mV to about +30 mV. Click **Record Data** to display your results in the grid (and record your results in Chart 3).

? PREDICT Question 1
How will the action potential at R1 (or R2) change as you continue to increase the stimulus voltage?

_____

4. You will now continue to observe the effects of incremental increases of the stimulus voltage. Increase the voltage by 10 mV by clicking the **+** button beside the voltage display and then click **Single Stimulus.** Repeat this step until you reach the maximum voltage the stimulator can deliver.

Repeat this step until you stimulate the axon at 50 mV and then click **Record Data** to display your results in the grid (and record your results in Chart 3).

After you complete the experiment, take the online **Post-lab Quiz** for Activity 3.

Activity Questions

1. Explain why the threshold voltage is not always the same value (between axons and within an axon).

_____

_____

2. Describe how the action potential is regenerated by local ion flux at each location on the axon.

_____

_____

3. Why doesn't the peak value of the action potential increase with stronger stimuli?

_____

_____

The Action Potential: Importance of Voltage-Gated Na⁺ Channels

OBJECTIVES

1. To define the term *voltage-gated channel.*
2. To describe the effect of tetrodotoxin on the voltage-gated Na^+ channel.
3. To describe the effect of lidocaine on the voltage-gated Na^+ channel.
4. To examine the effects of tetrodotoxin and lidocaine on the action potential.
5. To predict the effect of lidocaine on pain perception and to predict the site of action in the sensory neurons (nociceptors) that sense pain.

Introduction

The action potential (as seen in Activity 3) is generated when voltage-gated sodium channels open in sufficient numbers. **Voltage-gated sodium channels** open when the membrane depolarizes. Each sodium channel that opens allows Na^+ ions to diffuse into the cell down their electrochemical gradient. When enough sodium channels open so that the amount of sodium ions that enters via these voltage-gated channels overcomes the leak of potassium ions (recall that the potassium leak via passive channels establishes and maintains the negative resting membrane potential), threshold for the action potential is reached, and an action potential is generated.

In this activity you will observe what happens when these voltage-gated sodium channels are blocked with chemicals. One such chemical is tetrodotoxin (TTX), a toxin found in puffer fish, which is extremely poisonous. Another such chemical is lidocaine, which is typically used to block pain in dentistry and minor surgery.

EQUIPMENT USED The following equipment will be depicted on-screen: nerve chamber; axon; oscilloscope—used to observe timing of stimuli and voltage changes in the axon; stimulator—used to set the stimulus voltage and the interval between stimuli and to deliver pulses that depolarize the axon; stimulation wires (S); recording electrodes (wires R1 and R2)—used to record voltage changes in the axon (The first set of recording electrodes, R1, is 2 centimeters from the stimulation wires, and the second set of recording electrodes, R2, is 2 centimeters from R1.); tetrodotoxin (TTX); lidocaine.

Experiment Instructions

Go to the home page in the PhysioEx software and click **Exercise 3: Neurophysiology of Nerve Impulses.** Click **Activity 4: The Action Potential: Importance of Voltage-Gated Na⁺ Channels,** and take the online **Pre-lab Quiz** for Activity 4.

After you take the online Pre-lab Quiz, click the **Experiment** tab and begin the experiment. The experiment instructions are reprinted here for your reference. The opening screen for the experiment is shown below.

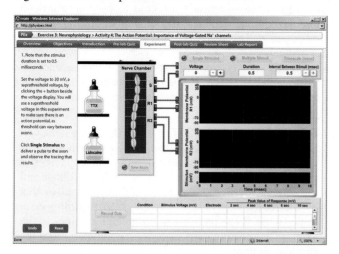

1. Note that the stimulus duration is set to 0.5 milliseconds. Set the voltage to 30 mV, a suprathreshold voltage, by clicking the + button beside the voltage display. You will use a suprathreshold voltage in this experiment to make sure there is an action potential, as threshold can vary between axons. Click **Single Stimulus** to deliver a pulse to the axon and observe the tracing that results.

2. Enter the peak value of the response at R1 and R2 in the field below and then click **Submit** to record your answer in the lab report. _____ μV

3. Click **Timescale** on the stimulator to change the timescale on the oscilloscope from milliseconds to seconds.

4. You will now deliver successive stimuli separated by 2.0-second intervals to observe what the control action potentials look like at this timescale. Set the interval between stimuli to 2.0 seconds by clicking the + button beside the "Interval between Stimuli" display. Click **Multiple Stimuli** to deliver pulses to the axon every 2 seconds. The stimuli will be stopped after 10 seconds.

5. Note the peak values of the responses at R1 and R2 and click **Record Data** to display your results in the grid (and record your results in Chart 4).

PREDICT Question 1
If you apply TTX between recording electrodes R1 and R2, what effect will the TTX have on the action potentials at R1 and R2?

_____

_____

6. Drag the dropper cap of the TTX bottle to the axon between recording electrodes R1 and R2 to apply a drop of TTX to the axon.

7. Click **Multiple Stimuli** to deliver pulses to the axon every 2 seconds. The stimuli will be stopped after 10 seconds.

8. Note the peak values of the responses at R1 and R2 and click **Record Data** to display your results in the grid (and record your results in Chart 4).

9. Click **New Axon** to select a new axon. TTX is irreversible and there is no known antidote for TTX poisoning.

PREDICT Question 2
If you apply lidocaine between recording electrodes R1 and R2, what effect will the lidocaine have on the action potentials at R1 and R2?

_____

_____

10. Drag the dropper cap of the lidocaine bottle to the axon between recording electrodes R1 and R2 to apply a drop of lidocaine to the axon.

11. Set the interval between stimuli to 2.0 seconds by clicking the + button beside the "Interval between Stimuli" display. Click **Multiple Stimuli** to deliver pulses to the axon every 2 seconds. The stimuli will be stopped after 10 seconds.

CHART 4	Effects of Tetrodotoxin and Lidocaine							
			Peak value of response (μV)					
Condition	Stimulus voltage (mV)	Electrodes	2 sec	4 sec	6 sec	8 sec	10 sec

12. Note the peak values of the responses at R1 and R2. For simplicity, this experiment was performed on a single axon, where the action potential is an "all-or-none" event. If you had treated a bundle of axons (a nerve), each with a slightly different threshold and sensitivity to the drugs, you would likely see the peak values of the action potentials decrease more gradually as more and more axons were blocked. Click **Record Data** to display your results in the grid (and record your results in Chart 4).

After you complete the experiment, take the online **Post-lab Quiz** for Activity 4.

Activity Questions

1. If depolarizing membrane potentials open voltage-gated sodium channels, what closes them?

_____

2. Why must a sushi chef go through years of training to prepare puffer fish for human consumption?

_____

3. For action potential generation and propagation, are there any other cation channels that could substitute for the voltage-gated sodium channels if the sodium channels were blocked?

_____

A C T I V I T Y   5

The Action Potential: Measuring Its Absolute and Relative Refractory Periods

OBJECTIVES

1. To define *inactivation* as it applies to a voltage-gated sodium channel.
2. To define the *absolute refractory period* and *relative refractory period* of an action potential.
3. To define the relationship between stimulus frequency and the generation of action potentials.

Introduction

Voltage-gated sodium channels in the plasma membrane of an excitable cell open when the membrane depolarizes. About 1–2 milliseconds later, these same channels inactivate, meaning they no longer allow sodium to go through the channel. These inactivated channels cannot be reopened by depolarization for an additional period of time (usually many milliseconds). Thus, during this time, fewer sodium channels can be opened. There are also voltage-gated potassium channels that open during the action potential. These potassium channels open more slowly. They contribute to the repolarization of the action potential from its peak, as more potassium flows out through this second type of potassium channel

(recall there are also passive potassium channels that let potassium leak out, and these leak channels are always open). The flux through extra voltage-gated potassium channels opposes the depolarization of the membrane to threshold, and it also causes the membrane potential to become transiently more negative than the resting potential at the end of an action potential. This phase is called after-hyperpolarization, or the undershoot.

In this activity you will explore what consequences the conformation states of voltage-gated channels have for the generation of subsequent action potentials.

> **EQUIPMENT USED** The following equipment will be depicted on-screen: nerve chamber; axon; oscilloscope—used to observe timing of stimuli and voltage changes in the axon; stimulator—used to set the stimulus voltage and the interval between stimuli and to deliver pulses that depolarize the axon; stimulation wires (S); recording electrode (wires R1)—used to record voltage changes in the axon. (The recording electrode is 2 centimeters from the stimulation wires.)

Experiment Instructions

Go to the home page in the PhysioEx software and click **Exercise 3: Neurophysiology of Nerve Impulses.** Click **Activity 5: The Action Potential: Measuring Its Absolute and Relative Refractory Periods,** and take the online **Pre-lab Quiz** for Activity 5.

After you take the online Pre-lab Quiz, click the **Experiment** tab and begin the experiment. The experiment instructions are reprinted here for your reference. The opening screen for the experiment is shown below.

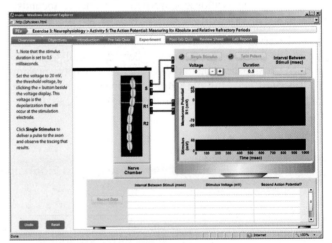

1. Note that the stimulus duration is set to 0.5 milliseconds. Set the voltage to 20 mV, the threshold voltage, by clicking the + button beside the voltage display. This voltage is the depolarization that will occur at the stimulation electrode. Click **Single Stimulus** to deliver a pulse to observe an action potential at this timescale.

2. You will now deliver two successive stimuli separated by 250 milliseconds. Set the interval between stimuli to 250 milliseconds by selecting 250 in the "Interval between Stimuli"

pull-down menu. Click **Twin Pulses** to deliver two pulses to the axon and observe the tracing that results. Click **Record Data** to display your results in the grid (and record your results in Chart 5).

CHART 5	Absolute and Relative Refractory Periods	
Interval between stimuli (msec)	Stimulus voltage (mV)	Second action potential?

3. Decrease the interval between stimuli to 125 milliseconds by selecting 125 in the "Interval between Stimuli" pull-down menu. Click **Twin Pulses** to deliver two pulses to the axon and observe the tracing that results. Click **Record Data** to display your results in the grid (and record your results in Chart 5).

4. Decrease the interval between stimuli to 60 milliseconds by selecting 60 in the "Interval between Stimuli" pull-down menu. Click **Twin Pulses** to deliver two pulses to the axon and observe the tracing that results.

Note that, at this stimulus interval, the second stimulus did not generate an action potential. Click **Record Data** to display your results in the grid (and record your results in Chart 5).

5. A second action potential can be generated at this stimulus interval, but the stimulus intensity must be increased. This interval is part of the relative refractory period, the time after an action potential when a second action potential can be generated if the stimulus intensity is increased.

Increase the stimulus intensity by 5 mV by clicking the + button beside the voltage display and then click **Twin Pulses** to deliver two pulses to the axon. Repeat this step until you generate a second action potential. After you generate a second action potential, click **Record Data** to display your results in the grid (and record your results in Chart 5).

? PREDICT Question 1
If you further decrease the interval between the stimuli, will the threshold for the second action potential change?

6. You will now decrease the interval until the second action potential fails again. (So that you can clearly observe two action potentials at the shorter interval between stimuli, the timescale on the oscilloscope has been set to 10 msec per division.) Decrease the interval between stimuli by 50% and then click **Twin Pulses** to deliver two pulses to the axon. When the second action potential fails, click **Record Data** to display your results in the grid (and record your results in Chart 5).

7. You will now increase the stimulus intensity until a second action potential is generated again. Increase the stimulus intensity by 5 mV by clicking the + button beside the voltage display and then click **Twin Pulses** to deliver two pulses to the axon. Repeat this step until you generate a second action potential. After you generate a second action potential, click **Record Data** to display your results in the grid (and record your results in Chart 5).

8. You will now determine the interval between stimuli at which a second action potential cannot be generated, no matter how intense the stimulus. Increase the stimulus intensity to 60 mV (the highest voltage on the stimulator). Decrease the interval between stimuli by 50% and then click **Twin Pulses** to deliver two pulses to the axon. Repeat this step until the second action potential fails.

The interval at which the second action potential fails is the **absolute refractory period,** the time after an action potential when the neuron cannot fire a second action potential, no matter how intense the stimulus. Click **Record Data** to display your results in the grid (and record your results in Chart 5).

After you complete the experiment, take the online **Post-lab Quiz** for Activity 5.

Activity Questions

1. Explain how the absolute refractory period ensures directionality of action potential propagation.

_____

_____

2. Some tissues (for example, cardiac muscle) have long absolute refractory periods. Why would this be beneficial?

_____

_____

3. What do you think is the benefit of a relative refractory period in an axon of a sensory neuron?

_____

_____

The Action Potential: Coding for Stimulus Intensity

OBJECTIVES

1. To observe the response of axons to longer periods of stimulation.
2. To examine the relationship between stimulus intensity and the frequency of action potentials.

Introduction

As seen in Activity 3, the action potential has a constant amplitude, regardless of the stimulus intensity—it is an "all-or-none" event. As seen in Activity 5, the absolute refractory period is the time after an action potential when the neuron cannot fire a second action potential, no matter how intense the stimulus, and the relative refractory period is the time after an action potential when a second action potential can be generated if the stimulus intensity is increased.

In this activity you will use these concepts to begin to explore how the axon codes the stimulus intensity as *frequency*, the number of events (in this case, action potentials) per unit time. To demonstrate this phenomenon you will use longer periods of stimulation that are more representative of real-life stimuli. For example, when you encounter an odor, the odor is normally present for seconds (or longer), unlike the very brief stimuli used in Activities 3–5. These longer stimuli allow the axon of the neuron to generate additional action potentials as soon as it has recovered from the first. As seen in Activity 5, the length of this recovery period changes depending on the stimulus intensity. For example, at threshold, a second action potential can occur only after the axon has recovered from the absolute refractory period and the entire relative refractory period.

We will not consider the phenomenon of adaptation, which is a decrease in the response amplitude that often occurs with prolonged stimuli. For example, with most odors, after many seconds, you no longer smell the odor, even though it is still present. This decrease in response is due to adaptation.

EQUIPMENT USED The following equipment will be depicted on-screen: nerve chamber; axon; oscilloscope—used to observe timing of stimuli and voltage changes in the axon; stimulator—used to set the voltage and duration of stimuli and to deliver pulses that depolarize the axon; stimulation wires (S); recording electrode (wires R1)—used to record voltage changes in the axon (The recording electrode is 2 centimeters from the stimulation wires).

Experiment Instructions

Go to the home page in the PhysioEx software and click **Exercise 3: Neurophysiology of Nerve Impulses.** Click **Activity 6: The Action Potential: Coding for Stimulus Intensity,** and take the online **Pre-lab Quiz** for Activity 6.

After you take the online Pre-lab Quiz, click the **Experiment** tab and begin the experiment. The experiment instructions are reprinted here for your reference. The opening screen for the experiment is shown below.

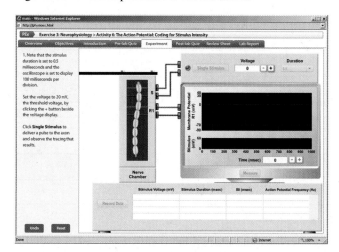

1. Note that the stimulus duration is set to 0.5 milliseconds and the oscilloscope is set to display 100 milliseconds per division. Set the voltage to 20 mV, the threshold voltage, by clicking the **+** button beside the voltage display. Click **Single Stimulus** to deliver a pulse to the axon and observe the tracing that results.

2. Note how the action potential looks at this timescale and click **Record Data** to display your results in the grid (and record your results in Chart 6).

CHART 6	Frequency of Action Potentials		
Stimulus voltage (mV)	Stimulus duration (msec)	ISI (msec)	Action potential frequency (Hz)

3. Increase the stimulus duration to 500 milliseconds by selecting 500 from the duration pull-down menu. Click **Single Stimulus** to deliver a pulse to the axon and observe the tracing that results. The stimulus is delivered after a delay of 100 milliseconds so that you can easily see the timing of the stimulus.

4. At the site of stimulation, the stimulus keeps the membrane of the axon at threshold for a long time, but this depolarization does not spread to the recording electrode. After one action potential has been generated and the axon has fully recovered from its absolute and relative refractory periods, the stimulus is still present to generate another action potential.

Measure the time (in milliseconds) between action potentials. This interval should be a bit longer than the relative refractory period (measured in Activity 5). Click **Measure** to help determine the time between action potentials. A thin, vertical yellow line appears at the far left side of the oscilloscope screen. You can move the line in 10-millisecond increments by clicking the + and – buttons beside the time display, which shows the time at the line. Click **Submit** to display your answer in the data table (and record your results in Chart 6).

5. The interval between action potentials is sometimes called the interspike interval (ISI). Action potentials are sometimes referred to as spikes because of their rapid time course. From the ISI, you can calculate the action potential frequency. The frequency is the reciprocal of the interval and is usually expressed in hertz (Hz), which is events (action potentials) per second. From the ISI you entered, calculate the frequency of action potentials with a prolonged (500 msec) threshold stimulus intensity. Frequency = 1/ISI. Click **Submit** to display your answer in the data table (and record your results in Chart 6).

6. A stimulus intensity of 30 mV was able to generate a second action potential toward the end of the relative refractory period in Activity 5. With this stronger stimulus, the second action potential can occur after a shorter time. Increase the stimulus intensity to 30 mV by clicking the + button beside the voltage display. Click **Single Stimulus** to deliver this stronger stimulus and observe the tracing that results.

7. Click **Submit** to display your answer in the data table (and record your results in Chart 6). Click **Measure** to help determine the time between action potentials. A thin, vertical yellow line appears at the far left side of the oscilloscope screen. You can move the line in 10-millisecond increments by clicking the + and – buttons beside the time display, which shows the time at the line.

8. From the ISI you entered, calculate the frequency of action potentials with a prolonged (500 msec) 30-mV stimulus intensity. Frequency = 1/ISI. Click **Submit** to display your answer in the data table (and record your results in Chart 6).

9. A stimulus intensity of 45 mV was able to generate a second action potential in the middle of the relative refractory period in Activity 5. With this even stronger stimulus, the second action potential can occur after an even shorter time. Increase the stimulus intensity to 45 mV.

? PREDICT Question 1
What effect will the increased stimulus intensity have on the frequency of action potentials?

10. Click **Single Stimulus** to deliver the stronger, 45-mV stimulus and observe the tracing that results.

11. Click **Submit** to display your answer in the data table (and record your results in Chart 6). Click **Measure** to help determine the time between action potentials. A thin, vertical yellow line appears at the far left side of the oscilloscope screen. You can move the line in 10-millisecond increments by clicking the + and – buttons beside the time display, which shows the time at the line.

12. From the ISI you entered, calculate the frequency of action potentials with a prolonged (500 msec) 45-mV stimulus intensity. Frequency = 1/ISI. Click **Submit** to display your answer in the data table (and record your results in Chart 6).

After you complete the experiment, take the online **Post-lab Quiz** for Activity 6.

Activity Questions

1. Compare the action potential frequency in a temperature-sensitive sensory neuron exposed to warm water and then hot water.

2. When a long-duration stimulus is applied, what two determinants of an action potential refractory period are being overcome?

3. Suggest several ways to pharmacologically overcome a neuron's refractory period and thereby increase the action potential frequency.

ACTIVITY 7

The Action Potential: Conduction Velocity

OBJECTIVES

1. To define and measure *conduction velocity* for an action potential.

2. To examine the effect of myelination on conduction velocity.

3. To examine the effect of axon diameter on conduction velocity.

Introduction

Once generated, the action potential is propagated, or conducted, down the axon. In other words, all-or-none action potentials are regenerated along the entire length of the axon. This propagation ensures that the amplitude of the action potential does not diminish as it is conducted along the axon. In some cases, such as the sensory neuron traveling from your toe to the spinal cord, the axon can be quite long (in this case, up to a 1 meter). Propagation/conduction occurs because there are voltage-gated sodium and potassium channels located along the axon and because the large depolarization that constitutes the action potential (once generated at the trigger zone) easily brings the next region of the axon to threshold. The **conduction velocity** can be easily calculated by knowing both the distance the action potential travels and the amount of time it takes. Velocity has the units of distance per time, typically meters/second. An experimental stimulus artifact (see Activity 3) provides a convenient marker of the stimulus time because it travels very quickly (for our purposes, instantaneously) along the axon.

Several parameters influence the conduction velocity in an axon, including the axon diameter and the amount of myelination. **Myelination** refers to a special wrapping of the membrane from glial cells (or neuroglia) around the axon. In the central nervous system, oligodendrocytes are the glia that wrap around the axon. In the peripheral nervous system, the Schwann cells are the glia that wrap around the axon. Many glial cells along the axon contribute a myelin sheath, and the myelin sheaths are separated by gaps called nodes of Ranvier.

In this activity you will compare the conduction velocities of three axons: (1) a large-diameter, heavily myelinated axon, often called an A fiber (the terms axon and fiber are synonymous), (2) a medium-diameter, lightly myelinated axon (called the B fiber), and (3) a thin, unmyelinated fiber (called the C fiber). Examples of these axon types in the body include the axon of the sensory Pacinian corpuscle (an A fiber), the axon of both the olfactory sensory neuron and a free nerve ending (C fibers), and a visceral sensory fiber (a B fiber).

EQUIPMENT USED The following equipment will be depicted on-screen: nerve chamber; three axons—A fiber, B fiber, and C fiber; oscilloscope—used to observe timing of stimuli and voltage changes in the axon; stimulator—used to set the stimulus voltage and to deliver pulses that depolarize the axon; stimulation wires (S); recording electrodes (wires R1 and R2)—used to record voltage changes in the axon. (The first set of recording electrodes, R1, is 2 centimeters from the stimulation wires, and the second set of recording electrodes, R2, is 2 centimeters from R1.)

Experiment Instructions

Go to the home page in the PhysioEx software and click **Exercise 3: Neurophysiology of Nerve Impulses.** Click **Activity 7: The Action Potential: Conduction Velocity,** and take the online **Pre-lab Quiz** for Activity 7.

After you take the online Pre-lab Quiz, click the **Experiment** tab and begin the experiment. The experiment instructions are reprinted here for your reference. The opening screen for the experiment is shown below.

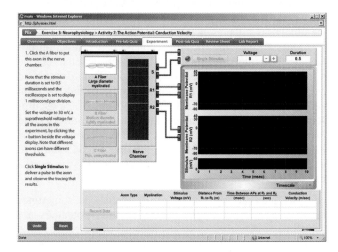

1. Click the A fiber to put this axon in the nerve chamber. Note that the stimulus duration is set to 0.5 milliseconds and the oscilloscope is set to display 1 millisecond per division.

Set the voltage to 30 mV, a suprathreshold voltage for all the axons in this experiment, by clicking the + button beside the voltage display. Note that different axons can have different thresholds. Click **Single Stimulus** to deliver a pulse to the axon and observe the tracing that results.

2. Click **Record Data** to display your results in the grid (and record your results in Chart 7).

3. Note the difference in time between the action potential recorded at R1 and the action potential recorded at R2. The distance between these sets of recording electrodes is 10 centimeters (0.1 m). Convert the time from milliseconds to seconds and then click **Submit** to display your results in the grid (and record your results in Chart 7).

4. Calculate the conduction velocity in meters/second by dividing the distance between R1 and R2 (0.1 m) by the time it took for the action potential to travel from R1 to R2. Click

---

CHART 7	Conduction Velocity					
Axon type	Myelination	Stimulus voltage (mV)	Distance from R1 to R2 (m)	Time between action potentials at R1 and R2		Conduction velocity (m/sec)
				(msec)	(sec)	

Submit to display your results in the grid (and record your results in Chart 7).

 PREDICT Question 1
How will the conduction velocity in the B fiber compare with that in the A fiber?

_____

_____

5. Click the **B fiber** to put this axon in the nerve chamber. Set the timescale on the oscilloscope to 10 milliseconds per division by selecting 10 in the timescale pull-down menu. Click **Single Stimulus** to deliver a pulse to the axon and observe the tracing that results.

6–8. Repeat steps 2–4 with the B fiber (and record your results in Chart 7).

 PREDICT Question 2
How will the conduction velocity in the C fiber compare with that in the B fiber?

_____

_____

9. Click the **C fiber** to put this axon in the nerve chamber. Set the timescale on the oscilloscope to 50 milliseconds per division by selecting 50 in the timescale pull-down menu. Click **Single Stimulus** to deliver a pulse to the axon and observe the tracing that results.

10–12. Repeat steps 2–4 with the C fiber (and record your results in Chart 7).

After you complete the experiment, take the online **Post-lab Quiz** for Activity 7.

Activity Questions

1. The squid utilizes a very large-diameter, unmyelinated axon to execute a rapid escape response when it perceives danger. How is this possible, given that the axon is unmyelinated?

_____

_____

2. When you burn your finger on a hot stove, you feel sharp, immediate pain, which later becomes slow, throbbing pain. These two types of pain are carried by different pain axons. Speculate on the axonal diameter and extent of myelination of these axons.

_____

_____

3. Why do humans possess a mixture of axons, some large-diameter, heavily myelinated axons and some small-diameter, relatively unmyelinated axons?

_____

_____

Chemical Synaptic Transmission and Neurotransmitter Release

OBJECTIVES

1. To define _neurotransmitter, chemical synapse, synaptic vesicle,_ and _postsynaptic potential._
2. To determine the role of calcium ions in neurotransmitter release.

Introduction

A major function of the nervous system is communication. The axon conducts the action potential from one place to another. Often, the axon has branches so that the action potential is conducted to several places at about the same time. At the end of each branch, there is a region called the axon terminal that is specialized to release packets of chemical neurotransmitters from small (~30-nm diameter) intracellular membrane-bound vesicles, called **synaptic vesicles.** **Neurotransmitters** are extracellular signal molecules that act on local targets as paracrine agents, on the neuron releasing the chemical as autocrine agents, and sometimes as hormones (endocrine agents) that reach their target(s) via the circulation. These chemicals are released by exocytosis and diffuse across a small extracellular space (called the synaptic gap, or synaptic cleft) to the target (most often the receiving end of another neuron or a muscle or gland). The neurotransmitter molecules often bind to membrane receptor proteins on the target, setting in motion a sequence of molecular events that can open or close membrane ion channels and cause the membrane potential in the target cell to change. This region where the neurotransmitter is released from one neuron and binds to a receptor on a target cell is called a **chemical synapse,** and the change in membrane potential of the target is called a synaptic potential, or **postsynaptic potential.**

In this activity you will explore some of the steps in neurotransmitter release from the axon terminal. Exocytosis of synaptic vesicles is normally triggered by an increase in calcium ions in the axon terminal. The calcium enters from outside the cell through membrane calcium channels that are opened by the depolarization of the action potential. The axon terminal has been greatly magnified in this activity so that you can visualize the release of neurotransmitter. Different from the other activities in this exercise, however, this procedure of directly seeing neurotransmitter release is not easily done in the lab; rather, neurotransmitter is usually detected by the postsynaptic potentials it triggers or by collecting and analyzing chemicals at the synapse after robust stimulation of the neurons.

EQUIPMENT USED The following equipment will be depicted on-screen: neuron (in vitro)—a large, dissociated (or cultured) neuron with magnified axon terminals; four extracellular solutions—control Ca^{2+}, no Ca^{2+}, low Ca^{2+}, and Mg^{2+}.

Experiment Instructions

Go to the home page in the PhysioEx software and click **Exercise 3: Neurophysiology of Nerve Impulses.** Click **Activity 8: Chemical Synaptic Transmission and Neurotransmitter Release,** and take the online **Pre-lab Quiz** for Activity 8.

After you take the online Pre-lab Quiz, click the **Experiment** tab and begin the experiment. The experiment instructions are reprinted here for your reference. The opening screen for the experiment is shown below.

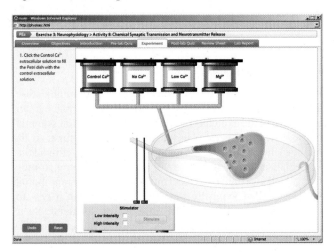

1. Click the *control Ca^{2+}* extracellular solution to fill the petri dish with the control extracellular solution.

2. Click **Low Intensity** on the stimulator and then click **Stimulate** to stimulate the neuron (axon) with a threshold stimulus that generates a low frequency of action potentials. Observe the release of neurotransmitter.

3. Click **High Intensity** on the stimulator and then click **Stimulate** to stimulate the neuron with a longer, more intense stimulus to generate a burst of action potentials. Observe the release of neurotransmitter.

? PREDICT Question 1
You have just observed that each action potential in a burst can trigger additional neurotransmitter release. If calcium ions are removed from the extracellular solution, what will happen to neurotransmitter release at the nerve terminal?

_____

_____

4–6. Repeat steps 1–3 with the *no Ca^{2+}* extracellular solution.

? PREDICT Question 2
What will happen to the amount of neurotransmitter release when low amounts of calcium are added back to the extracellular solution?

_____

_____

7–9. Repeat steps 1–3 with the *low Ca^{2+}* extracellular solution.

? PREDICT Question 3
What will happen to neurotransmitter release when magnesium is added to the extracellular solution?

_____

_____

10–12. Repeat steps 1–3 with the *Mg^{2+}* extracellular solution.

After you complete the experiment, take the online **Post-lab Quiz** for Activity 8.

Activity Questions

1. If you added more sodium to the extracellular solution, could the sodium substitute for the missing calcium?

_____

_____

2. How does botulinum toxin block synaptic transmission? Why is it used for cosmetic procedures?

_____

_____

A C T I V I T Y   9

The Action Potential: Putting It All Together

OBJECTIVES

1. To identify the functional areas (for example, the sensory ending, axon, and postsynaptic membrane) of a two-neuron circuit.
2. To predict and test the responses in each functional area to a very weak, subthreshold stimulus.
3. To predict and test the responses in each functional area to a moderate stimulus.
4. To predict and test the responses in each functional area to an intense stimulus.

Introduction

In the nervous system, sensory neurons respond to adequate sensory stimuli, generating action potentials in the axon if the stimulus is strong enough to reach threshold (the action potential is an "all-or-nothing" event). Via chemical synapses, these sensory neurons communicate with interneurons that process the information. Interneurons also communicate with motor neurons that stimulate muscles and glands, again, usually via chemical synapses.

After performing Activities 1–8, you should have a better understanding of how neurons function by generating changes from their resting membrane potential. If threshold is reached, an action potential is generated and propagated. If the stimulus is more intense, then action potentials are generated at a higher frequency, causing the release of more neurotransmitter at the next synapse. At an excitatory synapse the chemical neurotransmitter binds to receptors at the receiving end of the next cell (usually the cell body or dendrites of an interneuron), causing ion channels to open, resulting in a depolarization toward threshold for an action potential in the interneuron's axon. This depolarizing synaptic potential (called an excitatory postsynaptic potential) is graded in amplitude, depending on the amount of neurotransmitter and the number of channels that open. In the axon, the amplitude of this synaptic potential is coded as the frequency of action potentials. Neurotransmitters can also cause inhibition, which will not be covered in this activity.

In this activity you will stimulate a sensory neuron, predict the response of that cell and its target, and then test those predictions.

> **EQUIPMENT USED** The following equipment will be depicted on-screen: neuron (in vitro)—a large, dissociated (or cultured) neuron; interneuron (in vitro)—a large, dissociated (or cultured) interneuron; microelectrodes—small probes with very small tips that can impale a single neuron (In an actual wet lab, a microelectrode manipulator is used to position the microelectrodes. For simplicity, the microelectrode manipulator will not be depicted in this activity.); microelectrode amplifier—used to measure the voltage between the microelectrodes and a reference; oscilloscope—used to observe the changes in voltage across the membrane of the neuron and interneuron; stimulator—used to set the stimulus intensity (low or high) and to deliver pulses to the neuron.

Experiment Instructions

Go to the home page in the PhysioEx software and click **Exercise 3: Neurophysiology of Nerve Impulses.** Click **Activity 9: The Action Potential: Putting It All Together,** and take the online **Pre-lab Quiz** for Activity 9.

After you take the online Pre-lab Quiz, click the **Experiment** tab and begin the experiment. The experiment instructions are reprinted here for your reference. The opening screen for the experiment is shown below.

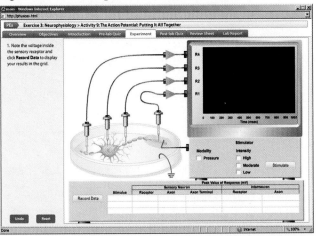

1. Note the membrane potential at the sensory receptor and the receiving end of the interneuron and click **Record Data** to display your results in the grid (and record your results in Chart 9).

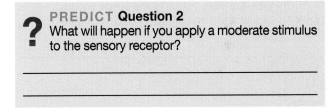

? PREDICT Question 1
What will happen if you apply a very weak, sub-threshold stimulus to the sensory receptor?

_____

_____

2. Click **Very Weak** intensity on the stimulator and then click **Stimulate** to stimulate the receiving end of the sensory neuron and observe the tracing that results.

3. Click **Record Data** to display your results in the grid (and record your results in Chart 9). The stimulus lasts 500 msec.

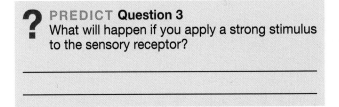

? PREDICT Question 2
What will happen if you apply a moderate stimulus to the sensory receptor?

_____

_____

4. Click **Moderate** intensity on the stimulator and then click **Stimulate** to stimulate the sensory receptor and observe the tracing that results.

5. Click **Record Data** to display your results in the grid (and record your results in Chart 9).

? PREDICT Question 3
What will happen if you apply a strong stimulus to the sensory receptor?

_____

_____

6. Click **Strong** intensity on the stimulator and then click **Stimulate** to stimulate the sensory receptor and observe the tracing that results.

7. Click **Record Data** to display your results in the grid (and record your results in Chart 9).

After you complete the experiment, take the online **Post-lab Quiz** for Activity 9.

CHART 9	Putting It All Together				
Stimulus	Sensory neuron			Interneuron	
	Membrane Potential (mV) Receptor	AP frequency (Hz) in axon	Vesicles released from axon terminal	Membrane potential (mV) receiving end	AP frequency (Hz) in axon
None					
Weak					
Moderate					
Strong					

Activity Questions

1. Why were the peak values of the action potentials at R2 and R4 the same when you applied a strong stimulus?

_____

_____

2. If the axons were unmyelinated, would the peak value of the action potential at R4 change relative to that at R2?

_____

_____

PhysioEx 9.0

Neurophysiology of Nerve Impulses

NAME_____

LAB TIME/DATE_____

The Resting Membrane Potential

1. Explain why increasing extracellular K^+ reduces the net diffusion of K^+ out of the neuron through the K^+ leak channels.

   _____

2. Explain why increasing extracellular K^+ causes the membrane potential to change to a less negative value. How well did the

   results compare with your prediction? _____

   _____

   _____

3. Explain why a change in extracellular Na^+ did not alter the membrane potential in the resting neuron. _____

   _____

4. Discuss the relative permeability of the membrane to Na^+ and K^+ in a resting neuron. _____

   _____

   _____

5. Discuss how a change in Na^+ or K^+ conductance would affect the resting membrane potential. _____

   _____

Receptor Potential

1. Sensory neurons have a resting potential based on the efflux of potassium ions (as demonstrated in Activity 1). What passive

   channels are likely found in the membrane of the olfactory receptor, in the membrane of the Pacinian corpuscle, and in the

   membrane of the free nerve ending? _____

   _____

   _____

2. What is meant by the term *graded potential*? _____

   _____

   _____

3. Identify which of the stimulus modalities induced the largest amplitude receptor potential in the Pacinian corpuscle. How well did the results compare with your prediction? _____

_____

4. Identify which of the stimulus modalities induced the largest-amplitude receptor potential in the olfactory receptors. How well did the results compare with your prediction? _____

_____

5. The olfactory receptor also contains a membrane protein that recognizes isoamyl acetate and, via several other molecules, transduces the odor stimulus into a receptor potential. Does the Pacinian corpuscle likely have this isoamyl acetate receptor protein? Does the free nerve ending likely have this isoamyl acetate receptor protein? _____

_____

6. What type of sensory neuron would likely respond to a green light? _____

_____

_____

ACTIVITY 3   The Action Potential: Threshold

1. Define the term *threshold* as it applies to an action potential. _____

_____

_____

2. What change in membrane potential (depolarization or hyperpolarization) triggers an action potential? _____

_____

3. How did the action potential at R1 (or R2) change as you increased the stimulus voltage above the threshold voltage? How well did the results compare with your prediction? _____

_____

4. An action potential is an "all-or-nothing" event. Explain what is meant by this phrase. _____

_____

_____

5. What part of a neuron was investigated in this activity? _____

_____

_____

ACTIVITY 4 The Action Potential: Importance of Voltage-Gated Na^+ Channels

1. What does TTX do to voltage-gated Na^+ channels? _____

_____

_____

2. What does lidocaine do to voltage-gated Na^+ channels? How does the effect of lidocaine differ from the effect of TTX?

_____

_____

3. A nerve is a bundle of axons, and some nerves are less sensitive to lidocaine. If a nerve, rather than an axon, had been used

in the lidocaine experiment, the responses recorded at R1 and R2 would be the sum of all the action potentials (called a compound

action potential). Would the response at R2 after lidocaine application necessarily be zero? Why or why not? _____

_____

_____

4. Why are fewer action potentials recorded at R2 when TTX is applied between R1 and R2? How well did the results compare

with your prediction? _____

_____

_____

5. Why are fewer action potentials recorded at R2 when lidocaine is applied between R1 and R2? How well did the results compare

with your prediction? _____

_____

_____

6. Pain-sensitive neurons (called nociceptors) conduct action potentials from the skin or teeth to sites in the brain involved in

pain perception. Where should a dentist inject the lidocaine to block pain perception? _____

_____

_____

ACTIVITY 5 The Action Potential: Measuring Its Absolute and Relative Refractory Periods

1. Define *inactivation* as it applies to a voltage-gated sodium channel. _____

_____

_____

2. Define the *absolute refractory period.* _____

_____

_____

3. How did the threshold for the second action potential change as you further decreased the interval between the stimuli?

How well did the results compare with your prediction? _____

_____

4. Why is it harder to generate a second action potential during the relative refractory period? _____

_____

_____

ACTIVITY 6   The Action Potential: Coding for Stimulus Intensity

1. Why are multiple action potentials generated in response to a long stimulus that is above threshold? _____

_____

_____

2. Why does the frequency of action potentials increase when the stimulus intensity increases? How well did the results compare

with your prediction? _____

_____

3. How does threshold change during the relative refractory period? _____

_____

_____

4. What is the relationship between the interspike interval and the frequency of action potentials? _____

_____

_____

ACTIVITY 7   The Action Potential: Conduction Velocity

1. How did the conduction velocity in the B fiber compare with that in the A fiber? How well did the results compare with your

prediction? _____

_____

_____

2. How did the conduction velocity in the C fiber compare with that in the B fiber? How well did the results compare with your

prediction? _____

_____

_____

3. What is the effect of axon diameter on conduction velocity? _____

_____

_____

4. What is the effect of the amount of myelination on conduction velocity? _____

_____

_____

5. Why did the time between the stimulation and the action potential at R1 differ for each axon? _____

_____

6. Why did you need to change the timescale on the oscilloscope for each axon? _____

_____

_____

A C T I V I T Y 8 Chemical Synaptic Transmission and Neurotransmitter Release

1. When the stimulus intensity is increased, what changes: the number of synaptic vesicles released or the amount of

neurotransmitter per vesicle? _____

_____

_____

2. What happened to the amount of neurotransmitter release when you switched from the control extracellular fluid to the

extracellular fluid with no Ca^{2+}? How well did the results compare with your prediction? _____

_____

_____

3. What happened to the amount of neurotransmitter release when you switched from the extracellular fluid with no Ca^{2+} to

the extracellular fluid with low Ca^{2+}? How well did the results compare with your prediction? _____

_____

_____

4. How did neurotransmitter release in the Mg^{2+} extracellular fluid compare to that in the control extracellular fluid? How well did the result compare with your prediction? _____

_____

_____

5. How does Mg^{2+} block the effect of extracellular calcium on neurotransmitter release? _____

_____

_____

ACTIVITY 9 The Action Potential: Putting It All Together

1. Why is the resting membrane potential the same value in both the sensory neuron and the interneuron? _____

_____

_____

2. Describe what happened when you applied a very weak stimulus to the sensory receptor. How well did the results compare with your prediction? _____

_____

_____

3. Describe what happened when you applied a moderate stimulus to the sensory receptor. How well did the results compare with your prediction? _____

_____

_____

4. Identify the type of membrane potential (graded receptor potential or action potential) that occurred at R1, R2, R3, and R4 when you applied a moderate stimulus. (View the response to the stimulus.) _____

_____

_____

5. Describe what happened when you applied a strong stimulus to the sensory receptor. How well did the results compare with your prediction? _____

_____

_____

Endocrine System Physiology

Exercise Overview

In the human body the **endocrine system** (in addition to the nervous system) coordinates and integrates the functions of different physiological systems (view Figure 4.1). Thus, the endocrine system plays a critical role in maintaining **homeostasis.** This role begins with chemicals, called **hormones,** secreted from ductless **endocrine glands,** which are tissues that have an epithelial origin. Endocrine glands secrete hormones into the extracellular fluid compartments. More specifically, the blood usually carries hormones (sometimes attached to specific plasma proteins) to their **target cells.** Target cells can be very close to, or very far from, the source of the hormone.

Hormones bind to high-affinity **receptors** located on the target cell's surface, in its cytosol, or in its nucleus. These hormone receptors have remarkable **sensitivity,** as the hormone concentration in the blood can range from 10^{-9} to 10^{-12} molar! A hormone-receptor complex forms and can then exert a **biological action** through signal-transduction cascades and alteration of gene transcription at the target cell. The physiological response to hormones can vary from seconds to hours to days, depending on the chemical nature of the hormone and its receptor location in the target cell.

The chemical structure of the hormone is important in determining how it will interact with target cells. *Peptide* and *catecholamine hormones* are fast-acting hormones that attach to a plasma-membrane receptor and cause a second-messenger cascade in the cytoplasm of the target cell. For example, a chemical called cAMP (cyclic adenosine monophosphate) is synthesized from a molecule of ATP. The

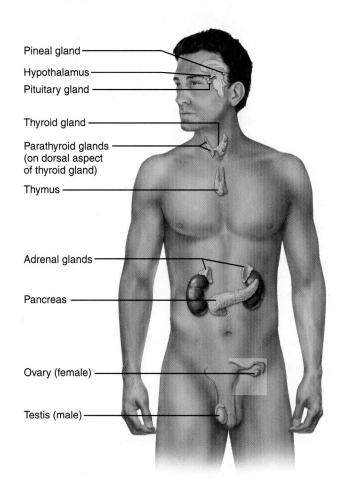

Pineal gland

Hypothalamus

Pituitary gland

Thyroid gland

Parathyroid glands
(on dorsal aspect
of thyroid gland)

Thymus

Adrenal glands

Pancreas

Ovary (female)

Testis (male)

FIGURE 4.1   Selected endocrine organs of the body.

synthesis of this chemical makes the cell more metabolically active and, therefore, more able to respond to a stimulus.

Steroid hormones and *thyroxine* (thyroid hormone) are slow-acting hormones that enter the target cell and interact with the nucleus to affect the transcription of various proteins that the cell can synthesize. The hormones enter the nucleus and attach at specific points on the DNA. Each attachment causes the production of a specific mRNA, which is then moved to the cytoplasm, where ribosomes can translate the mRNA into a protein.

Keep in mind that the organs of the endocrine system do not function independently. The activities of one endocrine gland are often coordinated with the activities of other glands. No one system functions independently of any other system. For this reason, we will be stressing feedback mechanisms and how we can use them to predict, explain, and understand hormone effects.

Given the powerful influence that hormones have on homeostasis, **negative feedback mechanisms** are important in regulating hormone secretion, synthesis, and effectiveness at target cells. Negative feedback ensures that if the body needs a particular hormone, that hormone will be produced until there is too much of it. When there is too much of the hormone, its release will be inhibited.

Rarely, the body regulates hormones via a *positive feedback mechanism.* The release of *oxytocin* from the posterior pituitary is one of these rare instances. Oxytocin is a hormone that causes the muscle layer of the uterus, called the *myometrium,* to contract during childbirth. This contraction of the myometrium causes additional oxytocin to be released,

allowing stronger contractions. Unlike what happens in negative feedback mechanisms, the increase in circulating levels of oxytocin does not inhibit oxytocin secretion.

Many experimental methods can be used to study the functions of an endocrine gland. These methods include removing the gland from an animal and then injecting, implanting, or feeding glandular extracts into a normal animal or an animal deprived of the gland being studied. In this exercise you will use these methods to gain a deeper understanding of the *function* and *regulation* of some of the endocrine glands.

ACTIVITY 1

Metabolism and Thyroid Hormone

OBJECTIVES

1. To understand the terms *basal metabolic rate (BMR), thyroid-stimulating hormone (TSH), thyroxine, goiter, hypothyroidism, hyperthyroidism, thyroidectomized,* and *hypophysectomized.*

2. To observe how negative feedback mechanisms regulate hormone release.

3. To understand thyroxine's role in maintaining the basal metabolic rate.

4. To understand the effect of TSH on the basal metabolic rate.

5. To understand the role of the hypothalamus in regulating the secretion of thyroxine and TSH.

Introduction

Metabolism is the broad range of biochemical reactions occurring in the body. Metabolism includes *anabolism* and *catabolism.* Anabolism is the building up of small molecules into larger, more complex molecules via enzymatic reactions. Energy is stored in the chemical bonds formed when larger, more complex molecules are formed.

Catabolism is the breakdown of large, complex molecules into smaller molecules via enzymatic reactions. The breaking of chemical bonds in catabolism releases energy that the cell can use to perform various activities, such as forming ATP. The cell does not use all the energy released by bond breaking. Much of the energy is released as heat to maintain a fixed body temperature, especially in humans. Humans are *homeothermic* organisms that need to maintain a fixed body temperature to maintain the activity of the various metabolic pathways in the body.

The most important hormone for maintaining metabolism and body heat is **thyroxine** (thyroid hormone), also known as *tetraiodothyronine,* or T_4. Thyroxine is secreted by the thyroid gland, located in the neck.

The production of thyroxine is controlled by the pituitary gland, or hypophysis, which secretes **thyroid-stimulating hormone (TSH).** The blood carries TSH to its target tissue, the thyroid gland. TSH causes the thyroid gland to increase in size and secrete thyroxine into the general circulation. If TSH levels are too high, the thyroid gland enlarges. The resulting glandular swelling in the neck is called a **goiter.**

The **hypothalamus** in the brain is also a vital participant in thyroxine and TSH production. It is a primary endocrine gland that secretes several hormones that affect the pituitary gland, or hypophysis, which is also located in the brain.

Thyrotropin-releasing hormone (TRH) is directly linked to thyroxine and TSH secretion. TRH from the hypothalamus stimulates the anterior pituitary to produce TSH, which then stimulates the thyroid to produce thyroxine.

These events are part of a classic negative feedback mechanism. When circulation levels of thyroxine are low, the hypothalamus secretes more TRH to stimulate the pituitary gland to secrete more TSH. The increase in TSH further stimulates the secretion of thyroxine from the thyroid gland. The increased levels of thyroxine will then influence the hypothalamus to reduce its production of TRH.

TRH travels from the hypothalamus to the pituitary gland via the **hypothalamic-pituitary portal system.** This specialized arrangement of blood vessels consists of a single **portal vein** that connects two capillary beds. The hypothalamic-pituitary portal system transports many other hormones from the hypothalamus to the pituitary gland. The hypothalamus primarily secretes *tropic* hormones, which stimulate the secretion of other hormones. TRH is an example of a tropic hormone because it stimulates the release of TSH from the pituitary gland. TSH itself is also an example of a tropic hormone because it stimulates production of thyroxine.

In this activity you will investigate the effects of thyroxine and TSH on a rat's metabolic rate. The metabolic rate will be indicated by the amount of oxygen the rat consumes per time per body mass. You will perform four experiments on three rats: a normal rat, a thyroidectomized rat (a rat whose thyroid gland has been surgically removed), and a hypophysectomized rat (a rat whose pituitary gland has been surgically removed). You will determine (1) the rat's basal metabolic rate, (2) its metabolic rate after it has been injected with thyroxine, (3) its metabolic rate after it has been injected with TSH, and (4) its metabolic rate after it has been injected with propylthiouracil, a drug that inhibits the production of thyroxine.

EQUIPMENT USED The following equipment will be depicted on-screen: three refillable syringes—used to inject the rats with propylthiouracil (a drug that inhibits the production of thyroxine by blocking the incorporation of iodine into the hormone precursor molecule), thyroid-stimulating hormone (TSH), and thyroxine; airtight, glass animal chamber—provides an isolated, sealed system in which to measure the amount of oxygen consumed by the rat in a specified amount of time (Opening the clamp on the left tube allows outside air into the chamber, and closing the clamp will create a closed, airtight system. The T-connector on the right tube allows you to connect the chamber to the manometer or to connect the fluid-filled manometer to the syringe filled with air.); soda lime (found at the bottom of the glass chamber)—absorbs the carbon dioxide given off by the rat; manometer—U-shaped tube containing fluid (As the rat consumes oxygen in the isolated, sealed system, this fluid will rise in the left side of the U-shaped tube and fall in the right side of the tube.); syringe—used to inject air into the tube and thus measure the amount of air that is needed to return the fluid columns in the manometer to their original levels; animal scale—used to measure body weight; three white rats—a *normal* rat, a *thyroidectomized* (Tx) rat (a rat whose thyroid gland has been surgically removed), and a *hypophysectomized* (Hypox) rat (a rat whose pituitary gland has been surgically removed).

Experiment Instructions

Go to the home page in the PhysioEx software and click **Exercise 4: Endocrine System Physiology.** Click **Activity 1: Metabolism and Thyroid Hormone,** and take the online **Pre-lab Quiz** for Activity 1.

After you take the online Pre-lab Quiz, click the **Experiment** tab and begin the experiment. The experiment instructions are reprinted here for your reference. The opening screen for the experiment is shown below.

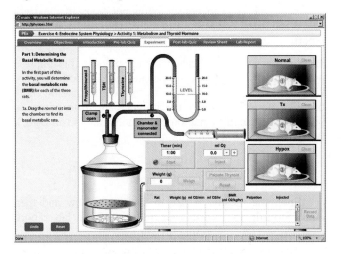

Part 1: Determining the Basal Metabolic Rates

In the first part of this activity, you will determine the basal metabolic rate (BMR) for each of the three rats.

1a. Drag the *normal* rat into the chamber to find its BMR.

1b. Click **Weigh** to determine the rat's weight.

1c. Click the clamp on the left tube (top of the chamber) to close it. This will prevent any outside air from entering the chamber and ensure that the only oxygen the rat is breathing is the oxygen inside the closed system.

1d. Note that the timer is set to one minute. Click **Start** beneath the timer to measure the amount of oxygen consumed by the rat in one minute in the sealed chamber. Note what happens to the water levels in the manometer as time progresses.

1e. Click the T-connector knob to connect the manometer and syringe.

1f. Click the clamp on the left tube (top of the chamber) to open it so the rat can breathe outside air.

1g. Observe the difference between the level in the left and right arms of the manometer. Estimate the volume of O_2 that you will need to inject to make the levels equal by counting the divisions on both sides. This volume is equivalent to the amount of oxygen that the rat consumed during the minute in the sealed chamber. Click the + button under the ml O_2 display until you reach the estimated volume. Then click **Inject** and watch what happens to the fluid in the two arms. When the volume levels are equalized, the word "Level" will appear and stay on the screen.

- If you have not injected enough oxygen, the word "Level" will not appear. Click the + to increase the volume and then click **Inject** again.

• If you have injected too much oxygen, the word "Level" will flash and then disappear. Click the button to decrease the volume and then click **Inject** again. Click **Record Data** when the levels are equalized.

1h. Calculate the oxygen consumption per hour for this rat using the following equation:

$$\frac{\text{ml O}_2 \text{ consumed}}{1 \text{ minute}} \times \frac{60 \text{ minutes}}{1 \text{ hr}} = \text{ml O}_2/\text{hr}$$

Enter the oxygen consumption per hour in the field below and then click **Submit** to record your results in the lab report. _____ ml O$_2$/hr

1i. Now that you have calculated the oxygen consumption per hour for this rat, you can calculate the metabolic rate per kilogram of body weight with the following equation (note that you need to convert the weight data from grams to kilograms to use this equation): Metabolic rate = (ml O$_2$/hr)/ (weight in kg) = ml O$_2$/kg/hr.

$$\text{Metabolic rate} = \frac{\text{ml O}_2/\text{hr}}{\text{weight in kg}} = \text{ml O}_2/\text{kg/hr}$$

Enter the metabolic rate in the field below and then click **Submit** to record your results in the lab report. _____ ml O$_2$/kg/hr

1j. Click **Palpate Thyroid** to manually check the size of the thyroid and, thus, whether a goiter is present. After reviewing the findings, click **Submit** to record your results in the lab report.

1k. Drag the rat from the chamber back to its cage and then click **Restore** (beneath **Palpate Thyroid**) to restore the apparatus to its initial state.

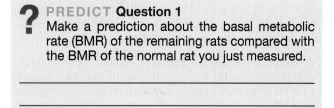

PREDICT Question 1
Make a prediction about the basal metabolic rate (BMR) of the remaining rats compared with the BMR of the normal rat you just measured.

_____

_____

2a.–2k. Repeat steps 1a–1k for the *thyroidectomized (Tx)* rat.

3a.–3k. Repeat steps 1a–1k for the *hypophysectomized (Hypox)* rat.

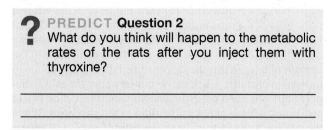

PREDICT Question 2
What do you think will happen to the metabolic rates of the rats after you inject them with thyroxine?

_____

_____

Part 2: Determining the Effect of Thyroxine on Metabolic Rate

In this part of the activity, you will investigate the effects of thyroxine injections on the metabolic rates of all three rats.

4a. Drag the syringe filled with *thyroxine* to the *normal* rat's hindquarters. Release the mouse button to inject thyroxine into the rat. (In this experiment, the effects of the injection are immediate. In a wet lab, you would have to inject the rats daily with thyroxine for 1–2 weeks).

4b. In this part of the activity, the rat's weight, the amount of oxygen consumed by the rat in one minute, the rat's oxygen consumption per hour, the rat's metabolic rate, and the result of the thyroid palpation will be generated automatically after you drag the rat into the chamber.

Drag the injected rat into the chamber and note the results (and record your results in Chart 1).

4c. Drag the rat from the chamber back to its cage and then click **Clean** to clear all traces of thyroxine from the rat and clean the syringe. (In this experiment, the thyroxine is removed instantly. In a wet lab, clearance would take weeks or require that a different rat be used.)

5a.–5c, Repeat steps 4a–4c with the *thyroidectomized (Tx)* rat (and record your results in Chart 1).

6a.–6c. Repeat steps 4a–4c with the *hypophysectomized (Hypox)* rat (and record your results in Chart 1).

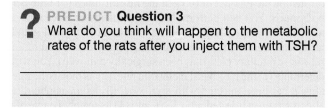

PREDICT Question 3
What do you think will happen to the metabolic rates of the rats after you inject them with TSH?

_____

_____

Part 3: Determining the Effect of TSH on Metabolic Rate

In this part of the activity, you will investigate the effects of TSH injections on the metabolic rates of all three rats.

7a. Drag the syringe filled with *TSH* to the *normal* rat's hindquarters. Release the mouse button to inject TSH into the rat. (In this experiment, the effects of the injection are immediate. In a wet lab, you would have to inject the rats daily with TSH for 1–2 weeks.)

7b. In this part of the activity, the rat's weight, the amount of oxygen consumed by the rat in one minute, the rat's oxygen consumption per hour, the rat's metabolic rate, and the result of the thyroid palpation will be generated automatically after you drag the rat into the chamber.

Drag the injected rat into the chamber and note the results (and record your results in Chart 1).

7c. Drag the rat from the chamber back to its cage and then click **Clean** to clear all traces of TSH from the rat and clean the syringe. (In this experiment, the TSH is removed instantly. In a wet lab, clearance would take weeks or require that a different rat be used.)

8a.–8c. Repeat steps 7a–7c with the *thyroidectomized (Tx)* rat (and record your results in Chart 1).

9a.–9c. Repeat steps 7a–7c with the *hypophysectomized (Hypox)* rat (and record your results in Chart 1).

CHART 1	Effects of Hormones on Metabolic Rate		
	Normal rat	**Thyroidectomized rat**	**Hypophysectomized rat**
Baseline			
Weight	_____ grams	_____ grams	_____ grams
ml O_2 used in 1 minute	_____ ml	_____ ml	_____ ml
ml O_2 used per hour	_____ ml	_____ ml	_____ ml
Metabolic rate	_____ ml O_2/kg/hr	_____ ml O_2/kg/hr	_____ ml O_2/kg/hr
Palpation results	_____	_____	_____
With thyroxine			
Weight	_____ grams	_____ grams	_____ grams
ml O_2 used in 1 minute	_____ ml	_____ ml	_____ ml
ml O_2 used per hour	_____ ml	_____ ml	_____ ml
Metabolic rate	_____ ml O_2/kg/hr	_____ ml O_2/kg/hr	_____ ml O_2/kg/hr
Palpation results	_____	_____	_____
With TSH			
Weight	_____ grams	_____ grams	_____ grams
ml O_2 used in 1 minute	_____ ml	_____ ml	_____ ml
ml O_2 used per hour	_____ ml	_____ ml	_____ ml
Metabolic rate	_____ ml O_2/kg/hr	_____ ml O_2/kg/hr	_____ ml O_2/kg/hr
Palpation results	_____	_____	_____
With propylthiouracil			
Weight	_____ grams	_____ grams	_____ grams
ml O_2 used in 1 minute	_____ ml	_____ ml	_____ ml
ml O_2 used per hour	_____ ml	_____ ml	_____ ml
Metabolic rate	_____ ml O_2/kg/hr	_____ ml O_2/kg/hr	_____ ml O_2/kg/hr
Palpation results	_____	_____	_____

? PREDICT Question 4
Propylthiouracil (PTU) is a drug that inhibits the production of thyroxine by blocking the attachment of iodine to tyrosine residues in the follicle cells of the thyroid gland (iodinated tyrosines are linked together to form thyroxine). What do you think will happen to the metabolic rates of the rats after you inject them with PTU?

_____

_____

Part 4: Determining the Effect of Propylthiouracil on Metabolic Rate

In this part of the activity, you will investigate the effects of propylthiouracil injections on the metabolic rates of all three rats.

10a. Drag the syringe filled with *propylthiouracil* to the *normal* rat's hindquarters. Release the mouse button to inject propylthiouracil into the rat. (In this experiment, the effects of the injection are immediate. In a wet lab, you would have to inject the rats daily with propylthiouracil for 1–2 weeks).

10b. In this part of the activity, the rat's weight, the amount of oxygen consumed by the rat in one minute, the rat's oxygen consumption per hour, the rat's metabolic rate, and the result of the thyroid palpation will be generated automatically after you drag the rat into the chamber.

Drag the injected rat into the chamber and note the results (and record your results in Chart 1).

10c. Drag the rat from the chamber back to its cage and then click **Clean** to clear all traces of propylthiouracil from the rat and clean the syringe. (In this experiment, the propylthiouracil is removed instantly. In a wet lab, clearance would take weeks or require that a different rat be used.)

11a.–11c. Repeat steps 10a–10c with the *thyroidectomized (Tx)* rat (and record your results in Chart 1).

12a.–12c. Repeat steps 10a–10c with the *hypophysectomized (Hypox)* rat (and record your results in Chart 1).

After you complete the experiment, take the online **Post-lab Quiz** for Activity 1.

Activity Questions

1. Using a water-filled manometer, you observed the amount of oxygen consumed by rats in a sealed chamber. What happened to the carbon dioxide the rat produced while in the sealed chamber?

_____

_____

2. What would happen to the fluid levels of the manometer (and, thus, the results of the metabolism experiment) if the rats in the sealed chamber were engaged in physical activity (such as running in a wheel)?

_____

_____

3. Describe the role of the hypothalamus in the production of thyroxine.

_____

_____

4. What does it mean if a hormone is a *tropic* hormone?

_____

_____

5. How could you treat a thyroidectomized rat so that it functions like a "normal" rat? How would you verify that your treatments were safe and effective?

_____

_____

6. What is the role of the hypothalamus in the production of thyroid-stimulating hormone (TSH)?

_____

_____

7. How does thyrotropin-releasing hormone (TRH) travel from the hypothalamus to the pituitary gland?

_____

_____

8. Why didn't the administration of TSH have any effect on the metabolic rate of the thyroidectomized rat?

_____

_____

9. Why didn't the administration of propylthiouracil have any effect on the metabolic rate of either the thyroidectomized rat or the hypophysectomized rat?

_____

_____

10. Propylthiouracil inhibits the production of thyroxine by blocking the attachment of iodine to the amino acid tyrosine. What naturally occurring problem in some parts of the world does this drug mimic?

_____

_____

Plasma Glucose, Insulin, and Diabetes Mellitus

OBJECTIVES

1. To understand the use of the terms *insulin, type 1 diabetes mellitus, type 2 diabetes mellitus,* and *glucose standard curve.*
2. To understand how fasting plasma glucose levels are used to diagnose diabetes mellitus.
3. To understand the assay that is used to measure plasma glucose.

Introduction

Insulin is a hormone produced by the beta cells of the endocrine portion of the pancreas. This hormone is vital to the regulation of **plasma glucose** levels, or "blood sugar," because the hormone enables our cells to absorb glucose from the bloodstream. Glucose absorbed from the blood is either used as fuel for metabolism or stored as glycogen (also known as animal starch), which is most notable in liver and muscle cells. About 75% of glucose consumed during a meal is stored as glycogen. As humans do not feed continuously (we are considered "discontinuous feeders"), the production of glycogen from a meal ensures that a supply of glucose will be available for several hours after a meal.

Furthermore, the body has to maintain a certain level of plasma glucose to continuously serve nerve cells because these cell types use only glucose for metabolic fuel. When glucose levels in the plasma fall below a certain value, the alpha cells of the pancreas are stimulated to release the hormone **glucagon.** Glucagon stimulates the breakdown of stored glycogen into glucose, which is then released back into the blood.

When the pancreas does not produce enough insulin, **type 1 diabetes mellitus** results. When the pancreas produces sufficient insulin but the body fails to respond to it, **type 2 diabetes mellitus** results. In either case, glucose remains in the bloodstream, and the body's cells are unable to take it up to serve as the primary fuel for metabolism. The kidneys then filter the excess glucose out of the plasma. Because the reabsorption of filtered glucose involves a finite number of transporters in kidney tubule cells, some of the excess glucose is not reabsorbed into the circulation. Instead, it passes out of the body in urine (hence *sweet urine,* as the name **diabetes mellitus** suggests).

The inability of body cells to take up glucose from the blood also results in skeletal muscle cells undergoing protein catabolism to free up amino acids to be used in forming glucose in the liver. This action puts the body into a negative nitrogen balance from the resulting protein depletion and tissue wasting. Other associated problems include poor wound healing and poor resistance to infections.

This activity is divided into two parts. In Part 1, you will generate a **glucose standard curve,** which will be explained in the experiment. In Part 2, you will use the glucose standard curve to measure the fasting plasma glucose levels from several patients to diagnose the presence or absence of diabetes mellitus. A patient with FPG values greater than or equal to 126 mg/dl in two FPG tests is diagnosed with diabetes. FPG values between 110 and 126 mg/dl indicate impairment or borderline impairment of insulin-mediated glucose uptake by cells. FPG values less than 110 mg/dl are considered normal.

EQUIPMENT USED The following equipment will be depicted on-screen: deionized water—used to adjust the volume so that it is the same for each reaction; glucose standard; enzyme color reagent; barium hydroxide; heparin; blood samples from five patients; test tubes—used as reaction vessels for the various tests; test tube incubation unit—used to incubate, mix, and centrifuge the samples; spectrophotometer—used to measure the amount of light absorbed or transmitted by a pigmented solution.

Experiment Instructions

Go to the home page in the PhysioEx software and click **Exercise 4: Endocrine System Physiology.** Click **Activity 2: Plasma Glucose, Insulin, and Diabetes Mellitus,** and take the online **Pre-lab Quiz** for Activity 2.

After you take the online Pre-lab Quiz, click the **Experiment** tab and begin the experiment. The experiment instructions are reprinted here for your reference. The opening screen for the experiment is shown below.

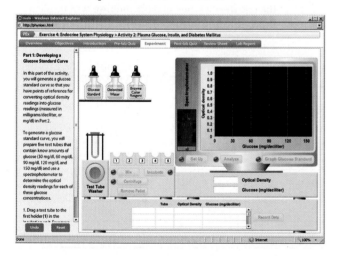

Part 1: Developing a Glucose Standard Curve

In this part of the activity, you will generate a glucose standard curve so that you have points of reference for converting optical density readings into glucose readings (measured in milligrams/deciliter, or mg/dl) in Part 2.

To generate a glucose standard curve, you will prepare five test tubes that contain known amounts of glucose (30 mg/dl, 60 mg/dl, 90 mg/dl, 120 mg/dl, and 150 mg/dl) and use a spectrophotometer to determine the optical density readings for each of these glucose concentrations.

1. Drag a test tube to the first holder (**1**) in the incubation unit. Four more test tubes will automatically be placed in the incubation unit.

2. Drag the dropper cap of the glucose standard bottle to the first tube in the incubation unit to dispense one drop of glucose standard solution into the tube. The dropper will automatically move across and dispense glucose standard to the remaining tubes. Note that each tube receives one additional drop of glucose standard (tube 2 receives 2 drops, tube 3 receives 3 drops, tube 4 receives 4 drops, and tube 5 receives 5 drops).

3. Drag the dropper cap of the deionized water bottle to the first tube in the incubation unit to dispense four drops of deionized water into the tube. The dropper will automatically move across and dispense deionized water to the remaining tubes. Note that each tube receives one less drop of deionized water (tube 2 receives 3 drops, tube 3 receives 2 drops, tube 4 receives 1 drop, and tube 5 does not receive any drops).

4. Click **Mix** to mix the contents of the tubes.

5. Click **Centrifuge** to centrifuge the contents of the tubes. After the centrifugation process, the tubes will automatically rise.

6. Click **Remove Pellet** to remove any pellets formed during the centrifugation process. Pellets can contain reagent precipitates and debris from the laboratory environment.

7. Drag the dropper cap of the enzyme color reagent bottle to the first tube in the incubation unit to dispense five drops of enzyme color reagent into each tube.

8. Click **Incubate** to incubate the contents of the tubes. The incubation unit will gently agitate the test tube rack, evenly mixing the contents of all test tubes throughout the incubation.

9. Click **Set Up** on the spectrophotometer to warm up the instrument and get it ready for your sample readings.

10. Drag tube 1 to the spectrophotometer.

11. Click **Analyze** to analyze the sample. A data point will appear on the monitor to show the optical density and the glucose concentration of the sample. These values will also appear in the optical density and glucose displays.

12. Click **Record Data** to display your results in the grid (**and** record your results in Chart 2.1). The tube will automatically be placed in the test tube washer.

CHART 2.1	Glucose Standard Curve Results	
Tube	**Optical density**	**Glucose (mg/dl)**
1		
2		
3		
4		
5		

13. You will now analyze the samples in the remaining tubes.

- Drag the next tube into the spectrophotometer.

- Click **Analyze** to analyze the sample. A data point will appear on the monitor to show the optical density and the glucose concentration of the sample. These values will also appear in the optical density and glucose displays.

- Click **Record Data** to display your results in the grid (and record your results in Chart 2.1). The tube will automatically be placed in the test tube washer.

Repeat this step until you analyze all five tubes.

14. Click **Graph Glucose Standard** to generate the glucose standard curve on the monitor. You will use this graph in Part 2.

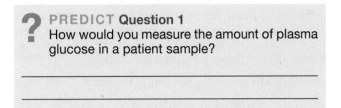

PREDICT Question 1
How would you measure the amount of plasma glucose in a patient sample?

_____

_____

Part 2: Measure Fasting Plasma Glucose Levels

In this part of the activity, you will use the glucose standard curve you generated in Part 1 to measure the fasting plasma glucose levels from five patients to diagnose the presence or absence of diabetes mellitus. Note the addition of two reagent bottles (barium hydroxide and heparin) and blood samples from the five patients. To undergo the fasting plasma glucose (FPG) test, patients must fast for a minimum of 8 hours prior to the blood draw.

A patient with FPG values greater than or equal to 126 mg/dl in two FPG tests is diagnosed with diabetes. FPG values between 110 and 126 mg/dl indicate impairment or borderline impairment of insulin-mediated glucose uptake by cells. FPG values less than 110 mg/dl are considered normal.

15. Drag a test tube to the first holder (**1**) in the incubation unit. Four more test tubes will automatically be placed in the incubation unit.

16. Drag the dropper cap of the first patient blood sample to the first tube in the incubation unit to dispense three drops of the sample. Three drops from each sample will automatically be dispensed into a separate tube.

17. Drag the dropper cap of the deionized water bottle to the first tube in the incubation unit to dispense five drops of deionized water into each tube.

18. Barium hydroxide dissolves and thus clears both proteins and cell membranes (so that clear glucose readings can be obtained). Drag the dropper cap of the barium hydroxide bottle to the first tube in the incubation unit to dispense five drops of barium hydroxide into each tube.

19. Drag the dropper cap of the heparin bottle to the first tube in the incubation unit to dispense a drop of heparin into each tube. Heparin prevents blood clots, which would interfere with clear glucose readings.

20. Click **Mix** to mix the contents of the tubes.

21. Click **Centrifuge** to centrifuge the contents of the tubes. After the centrifugation process, the tubes will automatically rise.

22. Click **Remove Pellet** to remove any pellets formed during the centrifugation process. Pellets can contain reagent precipitates and debris from the laboratory environment.

23. Drag the dropper cap of the enzyme color reagent bottle to the first tube in the incubation unit to dispense five drops of enzyme color reagent into each tube.

24. Click **Incubate** to incubate the contents of the tubes. The incubation unit will gently agitate the test tube rack, evenly mixing the contents of all test tubes throughout the incubation.

25. Click **Set Up** on the spectrophotometer to warm up the instrument and get it ready for your sample readings.

26. Click **Graph Glucose Standard** to display the glucose standard curve you generated in Part 1 on the monitor.

27. Drag tube 1 to the spectrophotometer.

28. Click **Analyze** to analyze the sample. A horizontal line will appear on the monitor to show the optical density of the sample. The optical density will also appear in the optical density display.

29. Drag the movable ruler (the vertical red line on the right side of the monitor) to the intersection of the horizontal yellow line (the optical density of the sample) and the glucose standard curve. Note the change in the glucose display as you move the line. The glucose concentration where the lines intersect is the fasting plasma glucose for this patient. Click **Record Data** to display your results in the grid (and record your results in Chart 2.2). The tube will automatically be placed in the test tube washer, and the monitor will be cleared (except for the glucose standard curve).

CHART 2.2	Fasting Plasma Glucose Results	
Sample	Optical density	Glucose (mg/dl)
1		
2		
3		
4		
5		

30. You will now analyze the samples in the remaining tubes.

- Drag the next tube into the spectrophotometer.

- Click **Analyze** to analyze the sample. A data point will appear on the monitor to show the optical density and the glucose concentration of the sample. These values will also appear in the optical density and glucose displays.

- Click **Record Data** to display your results in the grid. The tube will automatically be placed in the test tube washer (and record your results in Chart 2.2).

Repeat this step until you analyze all five tubes.

After you complete the experiment, take the online **Post-lab Quiz** for Activity 2.

Activity Questions

1. How would you know if your glucose standard curve was aberrant and thus inappropriate for patient diagnostics?

_____

_____

2. What are potential sources of variability when generating a glucose standard curve?

_____

_____

3. What recommendations would you make to a patient with fasting plasma glucose levels in the impaired/borderline-impaired range who was in the impaired/borderline-impaired range for the oral glucose tolerance test?

_____

_____

4. The amount of corn syrup in the American diet has been described as alarmingly high (especially in the foods that children eat). In the context of this activity, predict the likely trends in the fasting plasma glucose levels of our children as they mature.

_____

_____

Hormone Replacement Therapy

OBJECTIVES

1. To understand the terms *hormone replacement therapy, follicle-stimulating hormone (FSH), estrogen, calcitonin, osteoporosis, ovariectomized,* and *T score.*
2. To understand how estrogen levels affect bone density.
3. To understand the potential benefits of hormone replacement therapy.

Introduction

Follicle-stimulating hormone (FSH) is an anterior pituitary peptide hormone that stimulates ovarian follicle growth. Developing ovarian follicles then produce and secrete a steroid hormone called **estrogen** into the plasma. Estrogen has numerous effects on the female body and homeostasis, including the stimulation of bone growth and protection against **osteoporosis** (a reduction in the quantity of bone characterized by decreased bone mass and increased susceptibility to fractures).

After menopause, the ovaries stop producing and secreting estrogen. One of the effects and potential health problems of menopause is a loss of bone density that can result in osteoporosis and bone fractures. For this reason, postmenopausal treatments to prevent osteoporosis often include hormone replacement therapy. Estrogen can be administered to increase bone density. Calcitonin (secreted by C cells in the thyroid gland) is another peptide hormone that can be administered to counteract the development of osteoporosis. Calcitonin inhibits osteoclast activity and stimulates calcium uptake and deposition in long bones.

In this activity you will use three **ovariectomized** rats that are no longer producing estrogen because their ovaries have been surgically removed. A **T score** is a quantitative measurement of the mineral content of bone, used as an indicator of the structural strength of the bone and as a screen for osteoporosis. The three rats were chosen because each has a baseline T score of 2.61, indicating osteoporosis. T scores are interpreted as follows: normal = +1 to −0.99; osteopenia (bone thinning) = −1.0 to −2.49; osteoporosis = −2.5 and below.

You will administer either estrogen therapy or calcitonin therapy to these rats, representing two types of **hormone replacement therapy.** The third rat will serve as an untreated control and receive daily injections of saline. The vertebral bone density (VBD) of each rat will be measured with dual X-ray absorptiometry (DXA) to obtain its T score after treatment.

> **EQUIPMENT USED** The following equipment will be depicted on-screen: three ovariectomized rats (Note that if this were an actual wet lab, the ovariectomies would have been performed on the rats a month before the experiment to ensure that no residual hormones remained in the rats' systems.); saline; estrogen; calcitonin; reusable syringe—used to inject the rats; anesthesia—used to immobilize the rats for the X-ray scanning; dual X-ray absorptiometry bone-density scanner (DXA)—used to measure vertebral bone density of the rats.

Experiment Instructions

Go to the home page in the PhysioEx software and click **Exercise 4: Endocrine System Physiology.** Click **Activity 3: Hormone Replacement Therapy,** and take the online **Pre-lab Quiz** for Activity 3.

After you take the online Pre-lab Quiz, click the **Experiment** tab and begin the experiment. The experiment instructions are reprinted here for your reference. The opening screen for the experiment is shown on the following page.

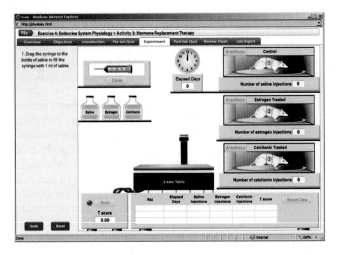

1. Drag the syringe to the bottle of saline to fill the syringe with 1 ml of saline.

2. Drag the syringe to the *control* rat, placing the tip of the needle in the rat's lower abdominal area. Injections into this area are considered *intraperitoneal* and will quickly be circulated by the abdominal blood vessels.

3. Click **Clean** beneath the syringe holder to clean the syringe of all residues.

4. Drag the syringe to the bottle of estrogen to fill the syringe with 1 ml of estrogen.

5. Drag the syringe to the *estrogen-treated* rat, placing the tip of the needle in the rat's lower abdominal area.

6. Click **Clean** beneath the syringe holder to clean the syringe of all residues.

7. Drag the syringe to the bottle of calcitonin to fill the syringe with 1 ml of calcitonin.

8. Drag the syringe to the *calcitonin-treated* rat, placing the tip of the needle in the rat's lower abdominal area.

9. Click **Clean** beneath the syringe holder to clean the syringe of all residues.

10. Click the clock face to advance one day (24 hours).

11. Each rat must receive seven injections over the course of seven days (one injection per day). The remaining injections will be automated. Click the clock face to repeat the series of injections until you have injected each of the rats seven times.

 PREDICT **Question 1**
What effect will the saline injections have on the control rat's vertebral bone density?

_____

_____

 PREDICT **Question 2**
What effect will the estrogen injections have on the estrogen-treated rat's vertebral bone density?

_____

_____

 PREDICT **Question 3**
What effect will the calcitonin injections have on the calcitonin-treated rat's vertebral bone density?

_____

_____

12. Click **Anesthesia** above the *control* rat's cage to immobilize the control rat with a gaseous anesthetic for X-ray scanning.

13. Drag the anesthetized rat to exam table for X-ray scanning.

14. Click **Scan** to activate the scanner. The T score will appear in the T score display. Click **Record Data** to record your results in the grid (and record your results in Chart 3). The control rat will be automatically returned to its cage.

CHART 3	Hormone Replacement Therapy Results
Rat	T score

15. You will now obtain the T scores for the remaining rats. Perform these steps to obtain the T score for the *estrogen-treated* rat, then repeat these steps to obtain the T score for the *calcitonin-treated* rat.

- Click **Anesthesia** above the rat's cage to immobilize the rat with a gaseous anesthetic for X-ray scanning.

- Drag the anesthetized rat to exam table for X-ray scanning.

- Click **Scan** to activate the scanner. The T score will appear in the T score display.

- Click **Record Data** to record your results in the grid (and record your results in Chart 3). The rat will be automatically returned to its cage.

After you complete the experiment, take the online **Post-lab Quiz** for Activity 3.

Activity Questions

1. Recently, hormone replacement therapy has been prominent in the popular press. Describe a hormone replacement therapy that you have seen in the news, and highlight its benefits, its potential risks, the reasons to continue and the reasons to discontinue its use.

_____

_____

2. In hormone replacement therapy, how is the hormone dose determined by the prescribing physician?

_____

_____ ▬

ACTIVITY 4

Measuring Cortisol and Adrenocorticotropic Hormone

OBJECTIVES

1. To understand the terms *cortisol, adrenocorticotropic hormone (ACTH), corticotropin-releasing hormone (CRH), Cushing's syndrome, iatrogenic, Cushing's disease,* and *Addison's disease.*

2. To understand how CRH controls ACTH secretion and ACTH controls cortisol secretion.

3. To understand how negative feedback mechanisms influence the levels of tropic CRH and ACTH.

4. To measure the blood levels of cortisol and ACTH in five patients and correlate these readings with symptoms and diagnoses.

5. To distinguish between Cushing's syndrome and Cushing's disease.

Introduction

Cortisol, a hormone secreted by the *adrenal cortex,* is important in the body's response to many kinds of stress. Cortisol release is stimulated by **adrenocorticotropic hormone (ACTH),** a tropic hormone released by the anterior pituitary. A *tropic* hormone stimulates the secretion of another hormone. ACTH release, in turn, is stimulated by **corticotropin-releasing hormone (CRH),** a tropic hormone from the hypothalamus. Increased levels of cortisol negatively feed back to inhibit the release of both ACTH and CRH.

Increased cortisol in the blood, or *hypercortisolism,* is referred to as **Cushing's syndrome** if the increase is caused by an adrenal gland tumor. Cushing's syndrome can also be **iatrogenic** (that is, physician induced). For example, physician-induced Cushing's syndrome can occur when glucocorticoid hormones, such as prednisone, are administered to treat rheumatoid arthritis, asthma, or lupus. Cushing's syndrome is often referred to as "steroid diabetes" because it results in hyperglycemia. In contrast, **Cushing's disease** is hypercortisolism caused by an anterior pituitary tumor. People with Cushing's disease exhibit increased levels of ACTH and cortisol.

Decreased cortisol in the blood, or *hypocortisolism,* can occur because of adrenal insufficiency. In primary adrenal insufficiency, also known as **Addison's disease,** the low cortisol is directly caused by gradual destruction of the adrenal cortex and ACTH levels are typically elevated as a compensatory effect. Secondary adrenal insufficiency also results in low levels of cortisol, usually caused by damage to the anterior pituitary. Therefore, the levels of ACTH are also low in secondary adrenal insufficiency.

As you can see, a variety of endocrine disorders can be related to both high and low levels of cortisol and ACTH. Table 4.1 summarizes these endocrine disorders.

TABLE 4.1	Cortisol and ACTH Disorders	
	Cortisol level	ACTH level
Cushing's syndrome (primary hypercortisolism)	High	Low
Iatrogenic Cushing's syndrome	High	Low
Cushing's disease (secondary hypercortisolism)	High	High
Addison's disease (primary adrenal insufficiency)	Low	High
Secondary adrenal insufficiency (hypopituitarism)	Low	Low

EQUIPMENT USED The following equipment will be depicted on-screen: plasma samples from five patients; HPLC (high-performance liquid chromatography) column— used to quantitatively measure the amount of cortisol and ACTH in the patient samples; HPLC detector—provides the hormone concentration in the patient sample; reusable syringe—used to inject the patient samples into the HPLC injection port; HPLC injection port—used to inject the patient samples into the HPLC column.

Experiment Instructions

Go to the home page in the PhysioEx software and click **Exercise 4: Endocrine System Physiology.** Click **Activity 4: Measuring Cortisol and Adrenocorticotropic Hormone,** and take the online **Pre-lab Quiz** for Activity 4.

After you take the online Pre-lab Quiz, click the **Experiment** tab and begin the experiment. The experiment instructions are reprinted here for your reference. The opening screen for the experiment is shown below.

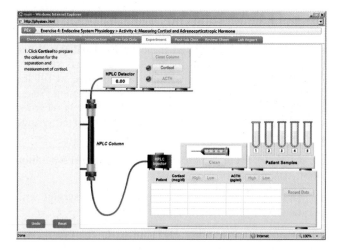

1. Click **Cortisol** to prepare the column for the separation and measurement of cortisol.

2. Drag the syringe to the first tube to fill the syringe with plasma isolated from the first patient.

3. Drag the syringe to the HPLC injector. The sample will enter the tubing and flow through the column. The cortisol concentration in the patient sample will appear in the HPLC detector display.

4. Click **Record Data** to display your results in the grid (and record your results in Chart 4).

CHART 4	Measurement of Cortisol			
Patient	Cortisol (mcg/dl)	Cortisol level	ACTH (pg/ml)	ACTH level
1				
2				
3				
4				
5				

5. Click **Clean** beneath the syringe to prepare it for the next sample. Click **Clean Column** to remove residual cortisol from the column.

6. Drag the syringe to the second tube to fill the syringe with plasma isolated from the second patient.

7. Drag the syringe to the HPLC injector. The sample will enter the tubing and flow through the column. The cortisol concentration in the patient sample will appear in the HPLC detector display.

8. Click **Record Data** to display your results in the grid (and record your results in Chart 4).

9. Click **Clean** beneath the syringe to prepare it for the next sample. Click **Clean Column** to remove residual cortisol from the column.

10. The procedure for the remaining samples will be completed automatically. Drag the syringe to the third tube to fill the syringe with plasma isolated from the third patient. When the cortisol concentration for the third patient is recorded in the grid, drag the syringe to the fourth tube to fill the syringe with plasma isolated from the fourth patient. When the cortisol concentration for the fourth patient is recorded in the grid, drag the syringe to the fifth tube to fill the syringe with plasma isolated from the fifth patient.

11. Click **ACTH** to prepare the column for ACTH separation and measurement.

12. Drag the syringe to the first tube to fill the syringe with plasma isolated from the first patient.

13. Drag the syringe to the HPLC injector. The sample will enter the tubing and flow through the column. The ACTH concentration in the patient sample will appear in the HPLC detector display.

14. Click **Record Data** to display your results in the grid.

15. Click **Clean** beneath the syringe to prepare it for the next sample. Click **Clean Column** to remove residual ACTH from the column.

16. Drag the syringe to the second tube to fill the syringe with plasma isolated from the second patient.

17. Drag the syringe to the HPLC injector. The sample will enter the tubing and flow through the column. The ACTH concentration in the patient sample will appear in the HPLC detector display.

18. Click **Record Data** to display your results in the grid (and record your results in Chart 4).

19. Click **Clean** beneath the syringe to prepare it for the next sample. Click **Clean Column** to remove residual ACTH from the column.

20. The procedure for the remaining samples will be completed automatically. Drag the syringe to the third tube to fill the syringe with plasma isolated from the third patient. When the ACTH concentration for the third patient is recorded in the grid, drag the syringe to the fourth tube to fill the syringe with plasma isolated from the fourth patient. When the ACTH concentration for the fourth patient is recorded in the grid, drag the syringe to the fifth tube to fill the syringe with plasma isolated from the fifth patient.

21. Indicate whether the cortisol and ACTH concentrations (levels) for each patient are high or low using the breakpoints shown in Table 4.2. Click the row of the patient and then click **High** or **Low** next to cortisol and ACTH.4.2.

TABLE 4.2	Abnormal Morning Cortisol and ACTH Levels	
ACTH level	High	Low
Cortisol	≥23 mcg/dl	<5 mcg/dl
ACTH	≥80 pg/ml	<20 pg/ml

Note: 1 mcg = 1 μg = 1 microgram

After you complete the experiment, take the online **Post-lab Quiz** for Activity 4.

Activity Questions

1. Discuss the benefits and drawbacks of giving glucocorticoids to young children that have significant allergy-induced asthma.

_____

_____

2. Explain the difference between Cushing's syndrome and Cushing's disease.

_____

_____

PhysioEx 9.0

REVIEW SHEET

EXERCISE 4

NAME ____________________

LAB TIME/DATE ____________________

Endocrine System Physiology

ACTIVITY 1 Metabolism and Thyroid Hormone

Part 1

1. Which rat had the fastest basal metabolic rate (BMR)? ____________________

2. Why did the metabolic rates differ between the normal rat and the surgically altered rats? How well did the results compare with your prediction? ____________________

3. If an animal has been thyroidectomized, what hormone(s) would be missing in its blood? ____________________

4. If an animal has been hypophysectomized, what effect would you expect to see in the hormone levels in its body? ____________________

Part 2

5. What was the effect of thyroxine injections on the normal rat's BMR? ____________________

6. What was the effect of thyroxine injections on the thyroidectomized rat's BMR? How does the BMR in this case compare with the normal rat's BMR? Was the dose of thyroxine in the syringe too large, too small, or just right? ____________________

7. What was the effect of thyroxine injections on the hypophysectomized rat's BMR? How does the BMR in this case compare

with the normal rat's BMR? Was the dose of thyroxine in the syringe too large, too small, or just right? —————————————

_____

_____

Part 3

8. What was the effect of thyroid-stimulating hormone (TSH) injections on the normal rat's BMR? _____

_____

_____

9. What was the effect of TSH injections on the thyroidectomized rat's BMR? How does the BMR in this case compare with

the normal rat's BMR? Why was this effect observed? _____

_____

10. What was the effect of TSH injections on the hypophysectomized rat's BMR? How does the BMR in this case compare with

the normal rat's BMR? Was the dose of TSH in the syringe too large, too small, or just right? _____

_____

_____

Part 4

11. What was the effect of propylthiouracil (PTU) injections on the normal rat's BMR? Why did this rat develop a palpable goiter?

_____

_____

12. What was the effect of PTU injections on the thyroidectomized rat's BMR? How does the BMR in this case compare with

the normal rat's BMR? Why was this effect observed? _____

_____

13. What was the effect of PTU injections on the hypophysectomized rat's BMR? How does the BMR in this case compare with

the normal rat's BMR? Why was this effect observed? _____

_____

_____

ACTIVITY 2 Plasma Glucose, Insulin, and Diabetes Mellitus

1. What is a glucose standard curve, and why did you need to obtain one for this experiment? Did you correctly predict how you would measure the amount of plasma glucose in a patient sample using the glucose standard curve? _____

_____

_____

2. Which patient(s) had glucose reading(s) in the diabetic range? Can you say with certainty whether each of these patients has type 1 or type 2 diabetes? Why or why not? _____

_____

_____

3. Describe the diagnosis for patient 3, who was also pregnant at the time of this assay. _____

_____

_____

4. Which patient(s) had normal glucose reading(s)? _____

_____

5. What are some lifestyle choices these patients with normal plasma glucose readings might recommend to the borderline impaired patients? _____

_____

_____

ACTIVITY 3 Hormone Replacement Therapy

1. Why were ovariectomized rats used in this experiment? How does the fact that the rats are ovariectomized explain their baseline T scores? _____

_____

_____

2. What effect did the administration of saline injections have on the control rat? How well did the results compare with your prediction? _____

_____

_____

3. What effect did the administration of estrogen injections have on the estrogen-treated rat? How well did the results compare

with your prediction? _____

_____

_____

4. What effect did the administration of calcitonin injections have on the calcitonin-treated rat? How well did the results

compare with your prediction? _____

_____

_____

5. What are some health risks that postmenopausal women must consider when contemplating estrogen hormone replacement

therapy? _____

_____

_____

ACTIVITY 4   **Measuring Cortisol and Adrenocorticotropic Hormone**

1. Which patient would most likely be diagnosed with Cushing's disease? Why? _____

_____

_____

2. Which two patients have hormone levels characteristic of Cushing's syndrome? _____

_____

_____

3. Patient 2 is being treated for rheumatoid arthritis with prednisone. How does this information change the diagnosis? _____

_____

_____

4. Which patient would most likely be diagnosed with Addison's disease? Why? _____

_____

_____

Cardiovascular Dynamics

Exercise Overview

The cardiovascular system is composed of a pump—the heart—and blood vessels that distribute blood containing oxygen and nutrients to every cell of the body. The principles governing blood flow are the same physical laws that apply to the flow of liquid through a system of pipes. For example, one very basic law in fluid mechanics is that the flow rate of a liquid through a pipe is directly proportional to the difference between the pressures at the two ends of the pipe (the **pressure gradient**) and inversely proportional to the pipe's **resistance** (a measure of the degree to which the pipe hinders, or resists, the flow of the liquid).

$$\text{Flow} = \text{pressure gradient/resistance} = \Delta P/R$$

This basic law also applies to blood flow. The "liquid" is blood, and the "pipes" are blood vessels. The pressure gradient is the difference between the pressure in arteries and the pressure in veins that results when blood is pumped into arteries. Blood flow rate is directly proportional to the pressure gradient and inversely proportional to resistance.

Blood flow is the amount of blood moving through a body area or the entire cardiovascular system in a given amount of time. Total blood flow is proportional to **cardiac output** (the amount of blood the heart is able to pump per minute). Blood flow to specific body areas can vary dramatically in a given time period. Organs differ in their requirements from moment to moment, and blood vessels have different sized diameters in their lumen (opening) to regulate local blood flow to various areas in response to the tissues' immediate needs. Consequently, blood flow can increase to some areas and decrease to other areas at the same time.

Resistance is a measure of the degree to which the blood vessel hinders, or resists, the flow of blood. The main factors that affect resistance are (1) blood vessel *radius,* (2) blood vessel *length,* and (3) blood *viscosity.*

Radius

The smaller the blood vessel radius, the greater the resistance, because of frictional drag between the blood and the vessel walls. Contraction of smooth muscle of the blood vessel, or **vasoconstriction,** results in a decrease in the blood vessel radius. Lipid deposits can also cause the radius of an artery to decrease, preventing blood from reaching the coronary tissue, which frequently leads to a heart attack. Alternately, relaxation of smooth muscle of the blood vessel, or **vasodilation,** causes an increase in the blood vessel radius. Blood vessel radius is the single most important factor in determining blood flow resistance.

Length

The longer the vessel length, the greater the resistance—again, because of friction between the blood and vessel walls. The length of a person's blood vessels change only as a person grows. Otherwise, the length generally remains constant.

Viscosity

Viscosity is blood "thickness," determined primarily by **hematocrit**—the fractional contribution of red blood cells to total blood volume. The higher the hematocrit, the greater the viscosity. Under most physiological conditions, hematocrit does not vary much and blood viscosity remains more or less constant.

The Effect of Blood Pressure and Vessel Resistance on Blood Flow

Blood flow is directly proportional to blood pressure because the pressure difference (ΔP) between the two ends of a vessel is the driving force for blood flow. Peripheral resistance is the friction that opposes blood flow through a blood vessel. This relationship is represented in the following equation:

$$\text{Blood flow (ml/min)} = \frac{\Delta P}{\text{peripheral resistance}}$$

Three factors that contribute to peripheral resistance are blood viscosity (η), blood vessel length *(L)*, and the radius of

the blood vessel *(r).* These relationships are expressed in the following equation:

$$\text{Peripheral resistance} = \frac{8L\eta}{\pi r^4}$$

From this equation you can see that the viscosity of the blood and the length of the blood vessel are directly proportional to peripheral resistance. The peripheral resistance is inversely proportional to the fourth power of the vessel radius. If you combine the two equations, you get the following result:

$$\text{Blood flow (ml/min)} = \frac{\Delta P \pi r^4}{8L\eta}$$

From this combination you can see that blood flow is directly proportional to the fourth power of vessel radius, which means that small changes in vessel radius result in dramatic changes in blood flow.

Studying the Effect of Blood Vessel Radius on Blood Flow Rate

OBJECTIVES

1. To understand how blood vessel radius affects blood flow rate.
2. To understand how vessel radius is changed in the body.
3. To understand how to interpret a graph of blood vessel radius versus blood flow rate.

Introduction

Controlling **blood vessel radius** (one-half of the diameter) is the principal method of controlling blood flow. Controlling blood vessel radius is accomplished by contracting or relaxing the smooth muscle within the blood vessel walls (vasoconstriction or vasodilation).

To understand why radius has such a pronounced effect on blood flow, consider the physical relationship between blood and the vessel wall. Blood in direct contact with the vessel wall flows relatively slowly because of the friction, or drag, between the blood and the lining of the vessel. In contrast, blood in the center of the vessel flows more freely because it is not rubbing against the vessel wall. The free-flowing blood in the middle of the vessel is called the **laminar flow.** Now picture a fully constricted (small-radius) vessel and a fully dilated (large-radius) vessel. In the fully constricted vessel, proportionately more blood is in contact with the vessel wall and there is less laminar flow, significantly impeding the rate of blood flow in the fully constricted vessel relative to that in the fully dilated vessel.

In this activity you will study the effect of blood vessel radius on blood flow. The experiment includes two glass beakers and a tube connecting them. Imagine that the left beaker is your heart, the tube is an artery, and the right beaker is a destination in your body, such as another organ.

EQUIPMENT USED The following equipment will be depicted on-screen: left beaker—simulates blood flowing from the heart; flow tube between the left and right beaker—simulates an artery; right beaker—simulates another organ (for example, the biceps brachii muscle).

Experiment Instructions

Go to the home page in the PhysioEx software and click **Exercise 5: Cardiovascular Dynamics.** Click **Activity 1: Studying the Effect of Blood Vessel Radius on Blood Flow Rate,** and take the online **Pre-lab Quiz** for Activity 1.

After you take the online Pre-lab Quiz, click the **Experiment** tab and begin the experiment. The experiment instructions are reprinted here for your reference. The opening screen for the experiment is shown below.

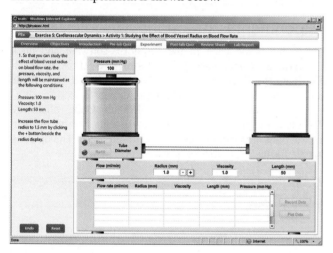

1. So that you can study the effect of blood vessel radius on blood flow rate, the pressure, viscosity, and length will be maintained at the following conditions:

Pressure: 100 mm Hg

Viscosity: 1.0

Length: 50 mm

Increase the flow tube radius to 1.5 mm by clicking the + button beside the radius display.

2. Click **Start** and then watch the fluid move into the right beaker. (Fluid moves slowly under some conditions—be patient!) Pressure propels fluid from the left beaker to the right beaker through the flow tube. The flow rate is shown in the flow rate display after the left beaker has finished draining.

3. Click **Record Data** to display your results in the grid (and record your results in Chart 1).

4. Click **Refill** to replenish the left beaker.

? PREDICT Question 1
What do you think will happen to the flow rate if the radius is increased by 0.5 mm?

_____

CHART 1	Effect of Blood Vessel Radius on Blood Flow Rate
Flow (ml/min)	Radius (mm)

5. Increase the flow tube radius to 2.0 mm by clicking the + button beside the radius display. Click **Start** and watch the fluid move into the right beaker.

6. Click **Record Data** to display your results in the grid (and record your results in Chart 1).

7. Click **Refill** to replenish the left beaker.

8. You will now observe the effect of incremental increases in flow tube radius.

- Increase the flow tube radius by 0.5 mm.
- Click **Start** and then watch the fluid move into the right beaker.
- Click **Record Data** to display your results in the grid (and record your results in Chart 1).
- Click **Refill** to replenish the left beaker.

Repeat this step until you reach a flow tube radius of 5.0 mm.

? PREDICT Question 2
Do you think a graph plotted with radius on the X-axis and flow rate on the Y-axis will be linear (a straight line)?

_____

9. Click **Plot Data** to view a summary of your data on a plotted grid. Radius will be displayed on the X-axis and flow rate will be displayed on the Y-axis. Click **Submit** to record your plot in the lab report.

After you complete the experiment, take the online **Post-lab Quiz** for Activity 1.

Activity Questions

1. Describe the relationship between vessel radius and blood flow rate.

_____

_____

2. In this activity you altered the radius of the flow tube by clicking the + and − buttons. Explain how and why the radius of blood vessels is altered in the human body.

_____

_____

3. Describe the appearance of your plot of blood vessel radius versus blood flow rate and relate the plot to the relationship between these two variables.

_____

_____

4. Describe an advantage of slower blood velocity in some areas of the body, for example, in the capillaries of our fingers.

_____

_____

ACTIVITY 2

Studying the Effect of Blood Viscosity on Blood Flow Rate

OBJECTIVES

1. To understand how blood viscosity affects blood flow rate.

2. To list the components in the blood that contribute to blood viscosity.

3. To explain conditions that might lead to viscosity changes in the blood.

4. To understand how to interpret a graph of viscosity versus blood flow.

Introduction

Viscosity is the thickness, or "stickiness," of a fluid. The more viscous a fluid, the more resistance to flow. Therefore, the flow rate will be slower for a more viscous solution. For example, consider how much more slowly maple syrup pours out of a container than milk does.

The viscosity of blood is due to the presence of plasma proteins and formed elements, which include white blood cells (leukocytes), red blood cells (erythrocytes), and platelets (thrombocytes). Formed elements and plasma proteins in the blood slide past one another, increasing the resistance to flow. With a viscosity of 3–5, blood is much more viscous than water (usually given a viscosity value of 1).

A body in homeostatic balance has a relatively stable blood consistency. Nevertheless, it is useful to examine the effects of blood viscosity on blood flow to predict what might occur in the human cardiovascular system when homeostatic imbalances occur. Factors such as dehydration and altered blood cell numbers do alter blood viscosity. For example,

polycythemia is a condition in which excess red blood cells are present, and certain types of anemia result in fewer red blood cells. Increasing the number of red blood cells increases blood viscosity, and decreasing the number of red blood cells decreases blood viscosity.

In this activity you will examine the effects of blood viscosity on blood flow rate. The experiment includes two glass beakers and a tube connecting them. Imagine that the left beaker is your heart, the tube is an artery, and the right beaker is a destination in your body, such as another organ.

> **EQUIPMENT USED** The following equipment will be depicted on-screen: left beaker—simulates blood flowing from the heart; flow tube between the left and right beaker—simulates an artery; right beaker—simulates another organ (for example, the biceps brachii muscle).

Experiment Instructions

Go to the home page in the PhysioEx software and click **Exercise 5: Cardiovascular Dynamics.** Click **Activity 2: Studying the Effect of Blood Viscosity on Blood Flow Rate,** and take the online **Pre-lab Quiz** for Activity 2.

After you take the online Pre-lab Quiz, click the **Experiment** tab and begin the experiment. The experiment instructions are reprinted here for your reference. The opening screen for the experiment is shown below.

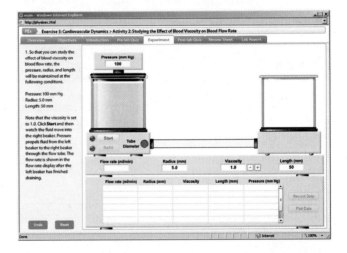

1. So that you can study the effect of blood viscosity on blood flow rate, the pressure, radius, and length will be maintained at the following conditions:

Pressure: 100 mm Hg

Radius: 5.0 mm

Length: 50 mm

Note that the viscosity is set to 1.0. Click **Start** and then watch the fluid move into the right beaker. Pressure propels fluid from the left beaker to the right beaker through the flow tube. The flow rate is shown in the flow rate display after the left beaker has finished draining.

2. Click **Record Data** to display your results in the grid (and record your results in Chart 2).

CHART 2	Effect of Blood Viscosity on Blood Flow Rate
Flow (ml/min)	Viscosity

3. Click **Refill** to replenish the left beaker.

> **? PREDICT Question 1**
> What effect do you think increasing the viscosity will have on the fluid flow rate?

4. Increase the fluid viscosity to 2.0 by clicking the **+** button beside the viscosity display. Click **Start** and then watch the fluid move into the right beaker.

5. Click **Record Data** to display your results in the grid (and record your results in Chart 2).

6. Click **Refill** to replenish the left beaker.

7. You will now observe the effect of incremental increases in viscosity.

- Increase the viscosity by 1.0.
- Click **Start** and then watch the fluid move into the right beaker.
- Click **Record Data** to display your results in the grid (and record your results in Chart 2).
- Click **Refill** to replenish the left beaker.

Repeat this step until you reach a viscosity of 8.0.

8. Click **Plot Data** to view a summary of your data on a plotted grid. Viscosity will be displayed on the X-axis and flow rate will be displayed on the Y-axis. Click **Submit** to record your plot in the lab report.

After you complete the experiment, take the online **Post-lab Quiz** for Activity 2.

Activity Questions

1. Describe the effect on blood flow rate when blood viscosity was increased.

2. Explain why the relationship between viscosity and blood flow rate is inversely proportional.

3. What might happen to blood flow if you increased the number of blood cells?

Studying the Effect of Blood Vessel Length on Blood Flow Rate

OBJECTIVES

1. To understand how blood vessel length affects blood flow rate.
2. To explain conditions that can lead to blood vessel length changes in the body.
3. To compare the effect of blood vessel length changes with the effect of blood vessel radius changes on blood flow rate.

Introduction

Blood vessel lengths increase as we grow to maturity. The longer the vessel, the greater the resistance to blood flow through the blood vessel because there is a larger surface area in contact with the blood cells. Therefore, when blood vessel length increases, friction increases. Our blood vessel lengths stay fairly constant in adulthood, unless we gain or lose weight. If we gain weight, blood vessel lengths can increase, and if we lose weight, blood vessel lengths can decrease.

In this activity you will study the physical relationship between blood vessel length and blood flow. Specifically, you will study how blood flow changes in blood vessels of constant radius but different lengths. The experiment includes two glass beakers and a tube connecting them. Imagine that the left beaker is your heart, the tube is an artery, and the right beaker is a destination in your body, such as another organ.

> **EQUIPMENT USED** The following equipment will be depicted on-screen: left beaker—simulates blood flowing from the heart; flow tube between the left and right beaker—simulates an artery; right beaker—simulates another organ (for example, the biceps brachii muscle).

Experiment Instructions

Go to the home page in the PhysioEx software and click **Exercise 5: Cardiovascular Dynamics.** Click **Activity 3: Studying the Effect of Blood Vessel Length on Blood Flow Rate,** and take the online **Pre-lab Quiz** for Activity 3.

After you take the online Pre-lab Quiz, click the **Experiment** tab and begin the experiment. The experiment instructions are reprinted here for your reference. The opening screen for the experiment is shown on the following page.

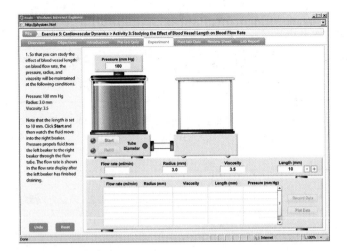

1. So that you can study the effect of blood vessel length on blood flow rate, the pressure, radius, and viscosity will be maintained at the following conditions:

Pressure: 100 mm Hg

Radius: 3.0 mm

Viscosity: 3.5

Note that the length is set to 10 mm. Click **Start** and then watch the fluid move into the right beaker. Pressure propels fluid from the left beaker to the right beaker through the flow tube. The flow rate is shown in the flow rate display after the left beaker has finished draining.

2. Click **Record Data** to display your results in the grid (and record your results in Chart 3).

CHART 3	Effect of Blood Vessel Length on Blood Flow Rate
Flow (ml/min)	**Flow Tube length (mm)**

3. Click **Refill** to replenish the left beaker.

? PREDICT Question 1
What effect do you think increasing the flow tube length will have on the fluid flow rate?

4. Increase the flow tube length to 15 mm by clicking the **+** button beside the length display. Click **Start** and then watch the fluid move into the right beaker.

5. Click **Record Data** to display your results in the grid (and record your results in Chart 3).

6. Click **Refill** to replenish the left beaker.

7. You will now observe the effect of incremental increases in flow tube length.

- Increase the flow tube length by 5 mm.
- Click **Start** and then watch the fluid move into the right beaker.
- Click **Record Data** to display your results in the grid (and record your results in Chart 3).
- Click **Refill** to replenish the left beaker.

  Repeat this step until you reach a flow tube length of 40 mm.

8. Click **Plot Data** to view a summary of your data on a plotted grid. Length will be displayed on the X-axis and flow rate will be displayed on the Y-axis. Click **Submit** to record your plot in the lab report.

After you complete the experiment, take the online **Post-lab Quiz** for Activity 3.

Activity Questions

1. Is the relationship between blood vessel length and blood flow rate directly proportional or inversely proportional? Why?

_____

_____

2. Which of the following can vary in size more quickly: blood vessel diameter or blood vessel length?

_____

_____

3. Describe what happens to resistance when blood vessel length increases.

_____

_____

ACTIVITY 4

Studying the Effect of Blood Pressure on Blood Flow Rate

OBJECTIVES

1. To understand how blood pressure affects blood flow rate.
2. To understand what structure produces blood pressure in the human body.
3. To compare the plot generated for pressure versus blood flow to those generated for radius, viscosity, and length.

Introduction

The pressure difference between the two ends of a blood vessel is the driving force behind blood flow. This pressure difference is referred to as a pressure gradient. In the cardiovascular system, the force of contraction of the heart provides the initial pressure and vascular resistance contributes to the pressure gradient. If the heart changes its force of contraction, the blood vessels need to be able to respond to the change in force. Large arteries close to the heart have more elastic tissue in their tunics in order to accommodate these changes.

In this activity you will look at the effect of pressure changes on blood flow (recall from the blood flow equation that a change in blood flow is directly proportional to the pressure gradient). The experiment includes two glass beakers and a tube connecting them. Imagine that the left beaker is your heart, the tube is an artery, and the right beaker is a destination in your body, such as another organ.

> **EQUIPMENT USED** The following equipment will be depicted on-screen: left beaker—simulates blood flowing from the heart; flow tube between the left and right beaker—simulates an artery; right beaker—simulates another organ (for example, the biceps brachii muscle).

Experiment Instructions

Go to the home page in the PhysioEx software and click **Exercise 5: Cardiovascular Dynamics.** Click **Activity 4: Studying the Effect of Blood Pressure on Blood Flow Rate,** and take the online **Pre-lab Quiz** for Activity 4.

After you take the online Pre-lab Quiz, click the **Experiment** tab and begin the experiment. The experiment instructions are reprinted here for your reference. The opening screen for the experiment is shown below.

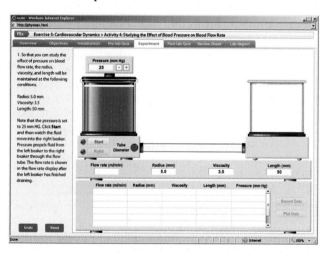

1. So that you can study the effect of pressure on blood flow rate, the radius, viscosity, and length will be maintained at the following conditions:

Radius: 5.0 mm

Viscosity: 3.5

Length: 50 mm

Note that the pressure is set to 25 mm Hg. Click **Start** and then watch the fluid move into the right beaker. Pressure propels fluid from the left beaker to the right beaker through the

flow tube. The flow rate is shown in the flow rate display after the left beaker has finished draining.

2. Click **Record Data** to record your results in the grid (and record your results in Chart 4).

CHART 4	Effect of Blood Pressure on Blood Flow Rate
Flow (ml/min)	Pressure (mm Hg)

3. Click **Refill** to replenish the left beaker.

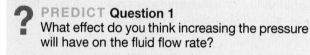

> **PREDICT Question 1**
> What effect do you think increasing the pressure will have on the fluid flow rate?

4. Increase the pressure to 50 mm Hg by clicking the **+** button beside the pressure display. Click **Start** and then watch the fluid move into the right beaker.

5. Click **Record Data** to record your results in the grid (and record your results in Chart 4).

6. Click **Refill** to replenish the left beaker.

7. You will now observe the effect of incremental increases in pressure.

- Increase the pressure by 25 mm Hg.

- Click **Start** and then watch the fluid move into the right beaker.

- Click **Record Data** to display your results in the grid (and record your results in Chart 4).

- Click **Refill** to replenish the left beaker.

Repeat this step until you reach a pressure of 200 mm Hg.

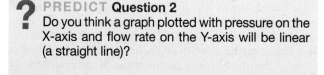

> **PREDICT Question 2**
> Do you think a graph plotted with pressure on the X-axis and flow rate on the Y-axis will be linear (a straight line)?

8. Click **Plot Data** to view a summary of your data on a plotted grid. Pressure will be displayed on the X-axis and flow rate will be displayed on the Y-axis. Click **Submit** to record your plot in the lab report.

After you complete the experiment, take the online **Post-lab Quiz** for Activity 4.

Activity Questions

1. How does increasing the driving pressure affect the blood flow rate?

_____

_____

2. Is the relationship between blood pressure and blood flow rate directly proportional or inversely proportional? Why?

_____

_____

3. How does the cardiovascular system increase pressure?

_____

_____

4. Although changing blood pressure can be used to alter the blood flow rate, this approach causes problems if it continues indefinitely. Explain why.

_____

_____

_____

ACTIVITY 5

Studying the Effect of Blood Vessel Radius on Pump Activity

OBJECTIVES

1. To understand the terms *systole* and *diastole*.
2. To predict how a change in blood vessel radius will affect flow rate.
3. To predict how a change in blood vessel radius will affect heart rate.
4. To observe the compensatory mechanisms for maintaining blood pressure.

Introduction

In the human body, the heart beats approximately 70 strokes each minute. Each heartbeat consists of a filling interval, when blood moves into the chambers of the heart, and an ejection period, when blood is actively pumped into the aorta and the pulmonary trunk.

The pumping activity of the heart can be described in terms of the phases of the cardiac cycle. Heart chambers fill during **diastole** (relaxation of the heart) and pump blood out during **systole** (contraction of the heart). As you can imagine, the length of time the heart is relaxed is one factor that determines the amount of blood within the heart at the end of the filling interval. Up to a point, increasing ventricular filling time results in a corresponding increase in ventricular volume. The volume in the ventricles at the end of diastole, just before cardiac contraction, is called the **end diastolic volume,** or **EDV.** The volume ejected by a single ventricular contraction is the **stroke volume,** and the volume remaining in the ventricle after contraction is the **end systolic volume,** or **ESV.**

The human heart is a complex, four-chambered organ consisting of two individual pumps (the right and left sides). The right side of the heart pumps blood through the lungs into the left side of the heart. The left side of the heart, in turn, delivers blood to the systems of the body. Blood then returns to the right side of the heart to complete the circuit.

Recall that cardiac output (**CO**) is equal to blood flow. To determine CO, you multiply heart rate (HR) by stroke volume (SV): CO = HR $\times$ SV. From the equation for flow (flow = $\Delta P/R$), you can determine the equation for blood pressure: ΔP = flow $\times$ R. Substituting CO in the equation for flow, you get: ΔP = HR $\times$ SV $\times$ R.

Therefore, to maintain blood pressure, the cardiovascular system can alter heart rate, stroke volume, or resistance. For example, if resistance decreases, heart rate can increase to maintain the pressure difference.

In this activity you will explore the operation of a simple, one-chambered pump and apply the physical concepts in the experiment to the operation of either of the two pumps of the human heart. The stroke volume and the difference in pressure will remain constant. You will explore the effect that a change in resistance has on heart rate and the compensatory mechanisms that the cardiovascular system uses to maintain blood pressure.

> **EQUIPMENT USED** The following equipment will be depicted on-screen: left beaker—simulates blood coming from the lungs; flow tube connecting the left beaker and the pump—simulates the pulmonary veins; pump—simulates the left ventricle (the valve to the left of the pump simulates the bicuspid valve, and the valve to the right of the pump simulates the aortic semilunar valve); flow tube connecting the pump and the right beaker—simulates the aorta; right beaker—simulates blood going to the systemic circuit.

Experiment Instructions

Go to the home page in the PhysioEx software and click **Exercise 5: Cardiovascular Dynamics.** Click **Activity 5: Compensation: Studying the Effect of Blood Vessel Radius on Pump Activity** and take the online **Pre-lab Quiz** for Activity 5.

After you take the online Pre-lab Quiz, click the **Experiment** tab and begin the experiment. The experiment instructions are reprinted here for your reference. The opening screen for the experiment is shown on the following page.

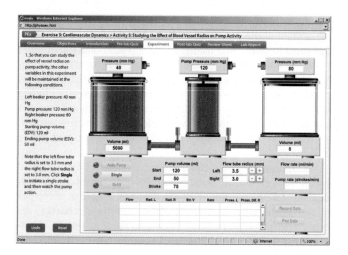

1. So that you can study the effect of vessel radius on pump activity, the other variables in this experiment will be maintained at the following conditions:

Left beaker pressure: 40 mm Hg

Pump pressure: 120 mm Hg

Right beaker pressure: 80 mm Hg

Starting pump volume (EDV): 120 ml

Ending pump volume (ESV): 50 ml

Note that the left flow tube radius is set to 3.5 mm and the right flow tube radius is set to 3.0 mm. Click **Single** to initiate a single stroke and then watch the pump action.

2. Click **Auto Pump** to initiate 10 strokes and then watch the pump action. The flow rate is shown in the flow rate display and the pump rate is shown in the pump rate display after the left beaker has finished draining.

3. Click **Record Data** to display your results in the grid (and record your results in Chart 5).

CHART 5	Effect of Blood Vessel Radius on Pump Activity	
Flow rate (ml/min)	Right radius (mm)	Pump rate (strokes/min)

4. Click **Refill** to replenish the left beaker.

PREDICT Question 1
If you increase the flow tube radius, what will happen to the pump rate to maintain constant pressure?

5. Increase the right flow tube radius to 3.5 mm by clicking the **+** button beside the right flow tube radius display. Click **Auto Pump** to initiate 10 strokes and then watch the pump action.

6. Click **Record Data** to display your results in the grid (and record your results in Chart 5).

7. Click **Refill** to replenish the left beaker.

8. You will now observe the effect of incremental increases in the right flow tube radius.

- Increase the right flow tube radius by 0.5 mm.
- Click **Auto Pump** to initiate 10 strokes and then watch the pump action.
- Click **Record Data** to display your results in the grid (and record your results in Chart 5).
- Click **Refill** to replenish the left beaker.

Repeat this step until you reach a right flow tube radius of 5.0 mm.

9. Click **Plot Data** to view a summary of your data on a plotted grid. Right flow tube radius will be displayed on the X-axis and flow rate will be displayed on the Y-axis. Click **Submit** to record your plot in the lab report.

After you complete the experiment, take the online **Post-lab Quiz** for Activity 5.

Activity Questions

1. Describe the position of the pump during diastole.

_____

_____

2. Describe the position of the pump during systole.

_____

_____

3. Describe what happened to the flow rate when the blood vessel radius was increased.

_____

_____

4. Explain what happened to the resistance and the pump rate to maintain pressure when the radius was increased.

_____

_____

Studying the Effect of Stroke Volume on Pump Activity

OBJECTIVES

1. To understand the effect a change in venous return has on stroke volume.
2. To explain how stroke volume is changed in the heart.
3. To explain the Frank-Starling law of the heart.
4. To define *preload, contractility,* and *afterload.*
5. To distinguish between intrinsic and extrinsic control of contractility of the heart.
6. To explore how heart rate and stroke volume contribute to cardiac output and blood flow.

Introduction

In a normal individual, 60% of the blood contained within the heart is ejected from the heart during ventricular systole, leaving 40% of the blood behind. The blood ejected by the heart—the **stroke volume**—is the difference between the **end diastolic volume (EDV),** the volume in the ventricles at the end of diastole, just before cardiac contraction, and **end systolic volume (ESV),** the volume remaining in the ventricle after contraction. That is, stroke volume = EDV − ESV. Many factors affect stroke volume, the most important of which include *preload, contractility,* and *afterload.* We will look at these defining factors and how they relate to stroke volume.

The Frank-Starling law of the heart states that, when more than the normal volume of blood is returned to the heart by the venous system, the heart muscle will be stretched, resulting in a more forceful contraction of the ventricles. This, in turn, will cause more than normal blood to be ejected by the heart, raising the stroke volume. The degree to which the ventricles are stretched by the end diastolic volume (EDV) is referred to as the **preload.** Thus, the preload results from the amount of ventricular filling between strokes, or the magnitude of the EDV. Ventricular filling could increase when the heart rate is slow because there will be more time for the ventricles to fill. Exercise increases venous return and, therefore, EDV. Factors such as severe blood loss and dehydration decrease venous return and EDV.

The **contractility** of the heart refers to strength of the cardiac muscle contraction (usually the ventricles) and its ability to generate force. A number of extrinsic mechanisms, including the sympathetic nervous system and hormones, control the force of cardiac muscle contraction, but they are not the focus of this activity. The focus of this activity will be the intrinsic controls of contractility (those that reside entirely within the heart). When the end diastolic volume increases, the cardiac muscle fibers of the ventricles stretch and lengthen. As the length of the cardiac sarcomere increases, so does the force of contraction. Cardiac muscle, like skeletal muscle, demonstrates a **length-tension relationship.** At rest, cardiac muscles are at a less than optimum overlap length for maximum tension production in the healthy heart. Therefore, when the heart experiences an increase in stretch with an increase in venous return and, therefore, EDV, it can respond by increasing the force of contraction, yielding a corresponding increase in stroke volume.

Afterload is the back pressure generated by the blood in the aorta and the pulmonary trunk. Afterload is the threshold that must be overcome for the aortic and pulmonary semilunar values to open. This pressure is referred to as an *after*load because the load is placed after the contraction of the ventricles starts. In the healthy heart, afterload doesn't greatly change stroke volume. However, individuals with high blood pressure can be affected because the ventricles are contracting against a greater pressure, possibly resulting in a decrease in stroke volume.

Cardiac output is equal to the heart rate (HR) multiplied by the stroke volume. Total blood flow is proportional to cardiac output (the amount of blood the heart is able to pump per minute). Therefore, when the stroke volume decreases, the heart rate must increase to maintain cardiac output. Conversely, when the stroke volume increases, the heart rate must decrease to maintain cardiac output.

Even though our simple pump in this experiment does not work exactly like the human heart, you can apply the concepts presented to basic cardiac function. In this activity you will examine how the activity of the pump is affected by changing the starting (EDV) and ending volumes (ESV).

> **EQUIPMENT USED** The following equipment will be depicted on-screen: left beaker—simulates blood coming from the lungs; flow tube connecting the left beaker and the pump—simulates the pulmonary veins; pump—simulates the left ventricle (the valve to the left of the pump simulates the bicuspid valve, and the valve to the right of the pump simulates the aortic semilunar valve); flow tube connecting the pump and the right beaker—simulates the aorta; right beaker—simulates blood going to the systemic circuit.

Experiment Instructions

Go to the home page in the PhysioEx software and click **Exercise 5: Cardiovascular Dynamics.** Click **Activity 6: Studying the Effect of Stroke Volume on Pump Activity** and take the online **Pre-lab Quiz** for Activity 6.

After you take the online Pre-lab Quiz, click the **Experiment** tab and begin the experiment. The experiment instructions are reprinted here for your reference. The opening screen for the experiment is shown below.

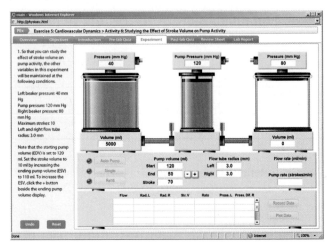

1. So that you can study the effect of stroke volume on pump activity, the other variables in this experiment will be maintained at the following conditions:

Left beaker pressure: 40 mm Hg

Pump pressure: 120 mm Hg

Right beaker pressure: 80 mm Hg

Maximum strokes: 10

Left and right flow tube radius: 3.0 mm

Note that the starting pump volume (EDV) is set to 120 ml. Set the stroke volume to 10 ml by increasing the ending pump volume (ESV) to 110 ml. To increase the ESV, click the + button beside the ending pump volume display.

2. Click **Auto Pump** to initiate 10 strokes and then watch the pump action. The flow rate is shown in the flow rate display and the pump rate is shown in the pump rate display after the left beaker has finished draining.

3. Click **Record Data** to display your results in the grid (and record your results in Chart 6).

CHART 6	Effect of Stroke Volume on Pump Activity	
Flow rate (ml/min)	Stroke volume (ml)	Pump rate (strokes/min)

4. Click **Refill** to replenish the left beaker.

PREDICT Question 1
If the pump rate is analogous to the heart rate, what do you think will happen to the rate when you increase the stroke volume?

5. Increase the stroke volume to 20 ml by decreasing the ESV. To decrease the ending pump volume, click the − button beside the ending pump volume display.

6. Click **Auto Pump** to initiate 10 strokes and then watch the pump action.

7. Click **Record Data** to display your results in the grid (and record your results in Chart 6).

8. Click **Refill** to replenish the left beaker.

9. You will now observe the effect of incremental increases in the stroke volume.

- Increase the stroke volume by 10 ml by decreasing the ending pump volume (ESV).
- Click **Auto Pump** to initiate 10 strokes and then watch the pump action.
- Click **Record Data** to display your results in the grid (and record your results in Chart 6).
- Click **Replenish** to refill the left beaker.

Repeat this step until you reach a stroke volume of 60 ml.

10. Increase the stroke volume by 20 ml by decreasing the ending pump volume (ESV). Click **Auto Pump** to initiate 10 strokes and then watch the pump action.

11. Click **Record Data** to display your results in the grid (and record your results in Chart 6).

12. Click **Refill** to replenish the left beaker.

13. Increase the stroke volume by 20 ml by decreasing the ending pump volume (ESV). Click **Auto Pump** to initiate 10 strokes and then watch the pump action.

14. Click **Record Data** to display your results in the grid (and record your results in Chart 6).

15. Click **Plot Data** to view a summary of your data on a plotted grid. Stroke volume will be displayed on the X-axis and flow rate will be displayed on the Y-axis. Click **Submit** to record your plot in the lab report.

After you complete the Experiment, take the online **Post-lab Quiz** for Activity 6.

Activity Questions

1. Describe how the heart responds to an increase in end diastolic volume (include the terms *preload* and *contractility* in your explanation).

2. Explain what happened to the pump rate when the stroke volume increased. Why?

3. Judging from the simulation results, explain why an athlete's resting heart rate might be lower than that of an average person.

Compensation in Pathological Cardiovascular Conditions

OBJECTIVES

1. To understand how aortic stenosis affects flow of blood through the heart.
2. To explain ways in which the cardiovascular system might compensate for changes in peripheral resistance.
3. To understand how the heart compensates for changes in afterload.
4. To explain how valves affect the flow of blood through the heart.

Introduction

If a blood vessel is compromised, your cardiovascular system can compensate to some degree. Aortic valve stenosis is a condition where there is a partial blockage of the aortic semilunar valve, increasing resistance to blood flow and left ventricular **afterload.** Therefore, the pressure that must be reached to open the aortic valve increases. The heart could compensate for a change in afterload by increasing contractility, the force of contraction. Increasing contractility will increase cardiac output by increasing stroke volume. To increase contractility, the myocardium becomes thicker. Athletes similarly improve their hearts through cardiovascular conditioning. That is, the thickness of the myocardium increases in diseased hearts with aortic valve stenosis and in athletes' hearts (though the chamber volume increases in athletes' hearts and decreases in diseased hearts).

Valves are important in the heart because they ensure that blood flows in one direction through the heart. The valves in the activity will ensure that blood moves in a single direction. Because the right flow tube represents the aorta (which is actually on the left side of the heart), decreasing the right flow tube radius simulates stenosis, or narrowing of the aortic valve.

Plaques in the arteries, known as **atherosclerosis,** can similarly cause an increase in resistance. An increase in peripheral resistance results in a decreased flow rate. Atherosclerosis is a type of **arteriosclerosis** in which the arteries have lost their elasticity. Atherosclerosis is one of the conditions that leads to heart disease.

In this activity you will test three different compensation mechanisms and predict which mechanism will make the best improvement in flow rate. The three mechanisms include (1) increasing the left flow tube radius (that is, increasing preload), (2) increasing the pump's pressure (that is, increasing contractility), and (3) decreasing the pressure in the right beaker (that is, decreasing afterload).

> **EQUIPMENT USED** The following equipment will be depicted on-screen: left beaker—simulates blood coming from the lungs; flow tube connecting the left beaker and the pump—simulates the pulmonary veins; pump—simulates the left ventricle (the valve to the left of the pump simulates the bicuspid valve, and the valve to the right of the pump simulates the aortic semilunar valve); flow tube connecting the pump and the right beaker—simulates the aorta; right beaker—simulates blood going to the systemic circuit.

Experiment Instructions

Go to the home page in the PhysioEx software and click **Exercise 5: Cardiovascular Dynamics.** Click **Activity 7: Compensation in Pathological Cardiovascular Conditions** and take the online **Pre-lab Quiz** for Activity 7.

After you take the online Pre-lab Quiz, click the **Experiment** tab and begin the experiment. The experiment instructions are reprinted here for your reference. The opening screen for the experiment is shown below.

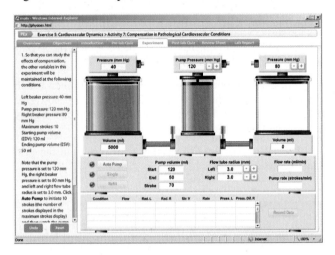

1. So that you can study the effects of compensation, the other variables in this experiment will be maintained at the following conditions:

Left beaker pressure: 40 mm Hg

Maximum strokes: 10

Starting pump volume (EDV): 120 ml

Ending pump volume (ESV): 50 ml

Note that the pump pressure is set to 120 mm Hg, the right beaker pressure is set to 80 mm Hg, and left and right flow tube radius is set to 3.0 mm. Click **Auto Pump** to initiate 10 strokes (the number of strokes displayed in the maximum strokes display) and then watch the pump action. The flow rate is shown in the flow rate display and the pump rate is shown in the pump rate display after the left beaker has finished draining.

2. Click **Record Data** to display your results in the grid (and record your results in Chart 7). This will be your baseline, or "normal," data point for flow rate.

3. Click **Refill** to replenish the left beaker.

4. Decrease the right flow tube radius to 2.5 mm by clicking the − button beside the right flow tube radius display. Click **Auto Pump** to initiate 10 strokes and then watch the pump action.

5. Click **Record Data** to display your results in the grid (and record your results in Chart 7).

6. Click **Refill** to replenish the left beaker.

CHART 7	Compensation Results						
Condition	Flow rate (ml/min)	Left radius (mm)	Right radius (mm)	Pump rate (strokes/ min)	Pump pressure (mm Hg)	Right beaker pressure (mm Hg)	

? PREDICT Question 1
You will now test three mechanisms to compensate for the decrease in flow rate caused by the decreased flow tube radius. Which mechanism do you think will have the greatest compensatory effect?

7. Increase the left flow tube radius to 3.5 mm by clicking the + button beside the left flow tube radius display. Click **Auto Pump** to initiate 10 strokes and then watch the pump action.

8. Click **Record Data** to display your results in the grid (and record your results in Chart 7).

9. Click **Refill** to replenish the left beaker.

10. Increase the left flow tube radius to 4.0 mm. Click **Auto Pump** to initiate 10 strokes and then watch the pump action.

11. Click **Record Data** to display your results in the grid (and record your results in Chart 7).

12. Click **Refill** to replenish the left beaker.

13. Increase the left flow tube radius to 4.5 mm. Click **Auto Pump** to initiate 10 strokes and then watch the pump action.

14. Click **Record Data** to display your results in the grid (and record your results in Chart 7).

15. Click **Refill** to replenish the left beaker.

16. Decrease the left flow tube radius to 3.0 mm by clicking the − button beside the left flow tube radius display and increase the pump pressure to 130 mm Hg by clicking the

+ button beside the pump pressure display. Click **Auto Pump** to initiate 10 strokes and then watch the pump action.

17. Click **Record Data** to display your results in the grid (and record your results in Chart 7).

18. Click **Refill** to replenish the left beaker.

19. Increase the pump pressure to 140 mm Hg by clicking the + button beside the pump pressure display. Click **Auto Pump** to initiate 10 strokes and then watch the pump action.

20. Click **Record Data** to display your results in the grid (and record your results in Chart 7).

21. Click **Refill** to replenish the left beaker.

22. Increase the pump pressure to 150 mm Hg. Click **Auto Pump** to initiate 10 strokes and then watch the pump action.

23. Click **Record Data** to display your results in the grid (and record your results in Chart 7).

24. Click **Refill** to replenish the left beaker.

25. Decrease the pump pressure to 120 mm Hg by clicking the − button beside the pump pressure display and decrease the right (destination) beaker pressure to 70 mm Hg by clicking the − button beside the right beaker pressure display. Click **Auto Pump** to initiate 10 strokes and then watch the pump action.

26. Click **Record Data** to display your results in the grid (and record your results in Chart 7).

27. Click **Refill** to replenish the left beaker.

28. Decrease the right (destination) beaker pressure to 60 mm Hg by clicking the − button beside the right beaker pressure display. Click **Auto Pump** to initiate 10 strokes and then watch the pump action.

29. Click **Record Data** to display your results in the grid (and record your results in Chart 7).

30. Click **Refill** to replenish the left beaker.

31. Decrease the right (destination) beaker pressure to 50 mm Hg. Click **Auto Pump** to initiate 10 strokes and then watch the pump action.

32. Click **Record Data** to display your results in the grid (and record your results in Chart 7).

33. Click **Refill** to replenish the left beaker.

> **?** **PREDICT Question 2**
> What do you think will happen if the pump pressure and the beaker pressure are the same?
> _____

34. Increase the right (destination) beaker pressure to 120 mm Hg by clicking the **+** button beside the right beaker pressure display. Click **Auto Pump** to initiate 10 strokes and then watch the pump action.

After you complete the experiment, take the online **Post-lab Quiz** for Activity 7.

Activity Questions

1. Explain why a thicker myocardium is seen in both the athlete's heart and the diseased heart.

_____

_____

2. Describe what the term *afterload* means.

_____

_____

3. Explain which mechanism in the simulation had the greatest compensatory effect.

_____

_____

4. Describe the mechanism used in the human heart to compensate for aortic stenosis.

_____

_____

Cardiovascular Dynamics

NAME_____

LAB TIME/DATE _____

ACTIVITY 1 Studying the Effect of Blood Vessel Radius on Blood Flow Rate

1. Explain how the body establishes a pressure gradient for fluid flow. _____

    _____

2. Explain the effect that the flow tube radius change had on flow rate. How well did the results compare with your prediction?

    _____

    _____

3. Describe the effect that radius changes have on the laminar flow of a fluid. _____

    _____

4. Why do you think the plot was not linear? (Hint: Look at the relationship of the variables in the equation.) How well did the

    results compare with your prediction? _____

    _____

ACTIVITY 2 Studying the Effect of Blood Viscosity on Blood Flow Rate

1. Describe the components in the blood that affect viscosity. _____

    _____

2. Explain the effect that the viscosity change had on flow rate. How well did the results compare with your prediction?

    _____

    _____

3. Describe the graph of flow versus viscosity. _____

    _____

4. Discuss the effect that polycythemia would have on viscosity and on blood flow. _____

    _____

    _____

ACTIVITY 3 Studying the Effect of Blood Vessel Length on Blood Flow Rate

1. Which is more likely to occur, a change in blood vessel radius or a change in blood vessel length? Explain why.

   _____

   _____

2. Explain the effect that the change in blood vessel length had on flow rate. How well did the results compare with your

   prediction? _____

   _____

3. Explain why you think blood vessel radius can have a larger effect on the body than changes in blood vessel length (use the

   blood flow equation). _____

   _____

4. Describe the effect that obesity would have on blood flow and why. _____

   _____

ACTIVITY 4 Studying the Effect of Blood Pressure on Blood Flow Rate

1. Explain the effect that pressure changes had on flow rate. How well did the results compare with your prediction?

   _____

2. How does the plot differ from the plots for tube radius, viscosity, and tube length? How well did the results compare with

   your prediction? _____

   _____

3. Explain why pressure changes are not the best way to control blood flow. _____

   _____

4. Use your data to calculate the increase in flow rate in ml/min/mm Hg. _____

   _____

ACTIVITY 5 Studying the Effect of Blood Vessel Radius on Pump Activity

1. Explain the effect of increasing the right flow tube radius on the flow rate, resistance, and pump rate. _____

   _____

2. Describe what the left and right beakers in the experiment correspond to in the human heart. _____

    _____

3. Briefly describe how the human heart could compensate for flow rate changes to maintain blood pressure. _____

    _____

ACTIVITY 6   Studying the Effect of Stroke Volume on Pump Activity

1. Describe the Frank-Starling law in the heart. _____

    _____

2. Explain what happened to the pump rate when you increased the stroke volume. Why do you think this occurred? How well

    did the results compare with your prediction? _____

    _____

3. Describe how the heart alters stroke volume. _____

    _____

4. Describe the intrinsic factors that control stroke volume. _____

    _____

ACTIVITY 7   Compensation in Pathological Cardiovascular Conditions

1. Explain how the heart could compensate for changes in peripheral resistance. _____

    _____

2. Which mechanism had the greatest compensatory effect? How well did the results compare with your prediction? _____

    _____

3. Explain what happened when the pump pressure and the beaker pressure were the same. How well did the results compare

    with your prediction? _____

    _____

4. Explain whether it would be better to adjust heart rate or blood vessel diameter to achieve blood flow changes at a local level

    (for example, in just the digestive system). _____

    _____

Cardiovascular Physiology

PRE-LAB QUIZ

1. Circle True or False: Cardiac muscle and some types of smooth muscle have the ability to contract without any external stimuli.

2. The total cardiac action potential lasts 250-300 milliseconds and has _____ phases.
   a. Two   c. Four
   b. Three   d. Five

3. The _____ nerve carries parasympathetic signals to the heart.
   a. Phrenic
   b. Vagus signals.
   c. Splanchnic
   d. Visceral

4. Circle True or False: Stimulation of the parasympathetic nervous system increases the heart rate and the force of contraction of the heart.

5. What is *vagal escape?* _____
   _____

6. The frog heart is different from the human heart in that it contains:
   a. One atria and two ventricles
   b. Two atria and a single, incompletely divided ventricle
   c. It is not different-it has two atria and two ventricles

7. The _____ node has the fastest rate of depolarization and determines the heart rate.
   a. Atrioventricular   c. Sinoatrial
   b. Atriosinus   d. Vetriculosino

8. Circle the correct term: Humans are able to maintain an external body temperature of 35.8-38.2° C in spite of environmental conditions and are referred to as poikilotherms / homeotherms.

9. Circle True or False: Sympathetic nerve fibers release epinephrine and acetylcholine at their cardiac synapses.

10. The resting cell membrane in cardiac muscle cells favors the movement of_____ ions.
    a. Potassium
    b. Calcium
    c. Sodium
    d. Magnesium

Exercise Overview

Cardiac muscle and some types of smooth muscle contract spontaneously, without any external stimuli. Skeletal muscle is unique in that it requires depolarizing signals from the nervous system to contract. The heart's ability to trigger its own contractions is called **autorhythmicity**.

If you isolate cardiac pacemaker muscle cells, place them into cell culture, and observe them under a microscope, you can see the cells contract. Autorhythmicity occurs because the plasma membrane in cardiac pacemaker muscle cells has reduced permeability to potassium ions but still allows sodium and calcium ions to slowly leak into the cells. This leakage causes the muscle cells to slowly depolarize until the action potential threshold is reached and L-type calcium channels open, allowing Ca^{2+} entry from the extracellular fluid. Shortly thereafter, contraction of the remaining cardiac muscle occurs prior to potassium-dependent repolarization. The spontaneous

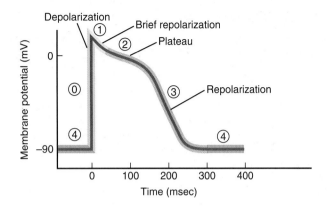

FIGURE 6.1 **The cardiac action potential.**

depolarization-repolarization events occur in a regular and continuous manner in cardiac pacemaker muscle cells, leading to **cardiac action potentials** in the majority of cardiac muscle.

There are five main phases of membrane polarization in a cardiac action potential (view Figure 6.1).

- **Phase 0** is similar to depolarization in the neuronal action potential. Depolarization causes voltage-gated sodium channels in the cell membrane to open, increasing the flow of sodium ions into the cell and increasing the membrane potential.

- In **phase 1**, the open sodium channels begin to inactivate, decreasing the flow of sodium ions into the cell and causing the membrane potential to fall slightly. At the same time, voltage-gated potassium channels close and voltage-gated calcium channels open. The subsequent decrease in the flow of potassium out of the cell and increase in the flow of calcium into the cell act to depolarize the membrane and curb the fall in membrane potential caused by the inactivation of sodium channels.

- In **phase 2**, known as the **plateau phase**, the membrane remains in a depolarized state. Potassium channels stay closed, and long-lasting (L-type) calcium channels stay open. This plateau lasts about 0.2 seconds, or 200 milliseconds.

- In **phase 3**, the membrane potential gradually falls to more negative values when a second set of potassium channels that began opening in phases 1 and 2 allows significant amounts of potassium to flow out of the cell. The falling membrane potential causes calcium channels to close, reducing the flow of calcium into the cell and repolarizing the membrane until the resting potential is reached.

- In **phase 4**, the resting membrane potential is again established in cardiac muscle cells and is maintained until the next depolarization arrives from neighboring cardiac pacemaker cells.

The total cardiac action potential lasts 250–300 milliseconds.

Investigating the Refractory Period of Cardiac Muscle

OBJECTIVES

1. To observe the autorhythmicity of the heart.
2. To understand the phases of the cardiac action potential.
3. To induce extrasystoles and observe them on the oscilloscope tracing of contractile activity in the isolated, intact frog heart.
4. To relate the presence or absence of wave summation and tetanus in cardiac muscle to the refractory period of the cardiac action potential.

Introduction

Recall that **wave summation** occurs when a skeletal muscle is stimulated with such frequency that muscle twitches overlap and result in a stronger contraction than a single muscle twitch. When the stimulations are frequent enough, the muscle reaches a state of fused tetanus, during which the individual muscle twitches cannot be distinguished. Tetanus occurs in skeletal muscle because skeletal muscle has a relatively short **absolute refractory period** (a period during which action potentials cannot be generated no matter how strong the stimulus).

Unlike skeletal muscle, cardiac muscle has a relatively long refractory period and is thus incapable of wave summation. In fact, cardiac muscle is incapable of reacting to *any* stimulus before approximately the middle of phase 3, and will not respond to a normal cardiac stimulus before phase 4. The period of time between the beginning of the cardiac action potential and the approximate middle of phase 3 is the **absolute refractory period**. The period of time between the absolute refractory period and phase 4 is the **relative refractory period**. The total refractory period of cardiac muscle is 200–250 milliseconds—almost as long as the contraction of the cardiac muscle.

In this activity you will use external stimulation to better understand the refractory period of cardiac muscle. You will use a frog heart, which is anatomically similar to the human heart. The frog heart has two atria and a single, incompletely divided ventricle.

EQUIPMENT USED The following equipment will be depicted on-screen: oscilloscope display—displays the contractile activity from the frog heart; electrical stimulator—used to a apply electrical shocks to the frog heart; electrode holder—locks electrodes in place for stimulation; external stimulation electrode; apparatus for sustaining an isolated frog heart—includes 23°C Ringer's solution; frog heart.

Experiment Instructions

Go to the home page in the PhysioEx software and click **Exercise 6: Cardiovascular Physiology.** Click **Activity 1: Investigating the Refractory Period of Cardiac Muscle,** and take the online **Pre-lab Quiz** for Activity 1.

After you take the online Pre-lab Quiz, click the **Experiment** tab and begin the experiment. The experiment instructions are reprinted here for your reference. The opening screen for the experiment is shown on the following page.

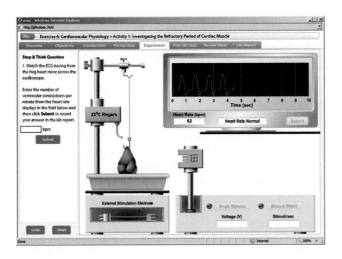

1. Watch the contractile activity from the frog heart on the oscilloscope. Enter the number of ventricular contractions per minute (from the heart rate display) in the field below and then click **Submit** to record your answer in the lab report.

_____ beats/min

2. Drag the external stimulation electrode to the electrode holder to the right of the frog heart. The electrode will touch the ventricular muscle tissue.

> ? PREDICT **Question 1**
> When you increase the frequency of the stimu-lation, what do you think will happen to the amplitude (height) of the ventricular systole wave?
> _____

3. Deliver single shocks in succession by clicking **Single Stimulus** rapidly. You might need to practice to acquire the correct technique. You should see a "doublet," or double peak, which contains an **extrasystole,** or extra contraction of the ventricle, and then a compensatory pause, which allows the heart to get back on schedule after the extrasystole. When you see a doublet, click **Submit** to record the tracing in the lab report.

> ? PREDICT **Question 2**
> If you deliver multiple stimuli (20 stimuli per second) to the heart, what do you think will happen?
> _____

4. Click **Multiple Stimuli** to deliver electrical shocks to the heart at a rate of 20 stimuli/sec. The **Multiple Stimuli** button changes to a **Stop Stimuli** button as soon as it is clicked. Observe the effects of stimulation on the contractile activity and, after a few seconds, click **Stop Stimuli** to stop the stimuli.

After you complete the experiment, take the online **Post-lab Quiz** for Activity 1.

Activity Questions

1. Describe how the frog heart and human heart differ anatomically.

_____

_____

2. What does an extrasystole correspond to? How did you induce an extrasystole on the ECG tracing?

_____

_____

3. Explain why it is important that wave summation and tetanus do not occur in the cardiac muscle.

_____

_____

ACTIVITY 2

Examining the Effect of Vagus Nerve Stimulation

OBJECTIVES

1. To understand the role that the sympathetic and parasympathetic nervous systems have on heart activity.
2. To explain the consequences of vagal stimulation and vagal escape.
3. To explain the functionality of the sinoatrial node.

Introduction

The autonomic nervous system has two branches: the **sympathetic** nervous system ("fight or flight") and **parasympathetic** nervous system ("resting and digesting"). At rest both the sympathetic and parasympathetic nervous systems are working but the parasympathetic branch is more active. The sympathetic nervous system becomes more active when needed, for example, during exercise and when confronting danger.

Both the parasympathetic and sympathetic nervous systems supply nerve impulses to the heart. Stimulation of the sympathetic nervous system increases the rate and force of contraction of the heart. Stimulation of the parasympathetic nervous system decreases the heart rate without directly changing the force of contraction. The vagus nerve (cranial nerve X) carries the signal to the heart. If stimulation of the vagus nerve (vagal stimulation) is excessive, the heart will stop beating. After a short time, the ventricles will begin to beat again. The resumption of the heartbeat is referred to as **vagal escape** and can be the result of sympathetic reflexes or initiation of a rhythm by the Purkinje fibers.

The **sinoatrial node (SA node)** is a cluster of autorhythmic cardiac cells found in the right atrial wall in the human heart. The SA node has the fastest rate of spontaneous depolarization, and, for that reason, it determines the heart rate and is therefore referred to as the heart's "**pacemaker**." In the absence of parasympathetic stimulation, sympathetic stimulation, and hormonal controls, the SA node generates action potentials 100 times per minute.

> **EQUIPMENT USED** The following equipment will be depicted on-screen: oscilloscope display—displays the contractile activity from the frog heart; electrical stimulator—used to apply electrical shocks to the frog heart; electrode holder—locks electrodes in place for stimulation; vagus nerve stimulation electrode; apparatus for sustaining an isolated, intact frog heart—includes 23°C Ringer's solution; frog heart with vagus nerve (thin, white strand to the right).

Experiment Instructions

Go to the home page in the PhysioEx software and click **Exercise 6: Cardiovascular Physiology.** Click **Activity 2: Examining the Effect of Vagus Nerve Stimulation,** and take the online **Pre-lab Quiz** for Activity 2.

After you take the online Pre-lab Quiz, click the **Experiment** tab and begin the experiment. The experiment instructions are reprinted here for your reference. The opening screen for the experiment is shown below.

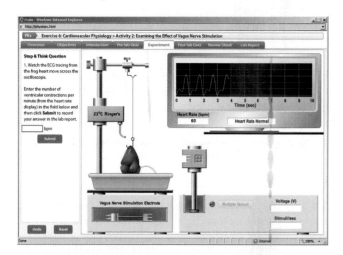

1. Watch the contractile activity from the frog heart on the oscilloscope. Enter the number of ventricular contractions per minute (from the heart rate display) in the field below and then click **Submit** to record your answer in the lab report.

_____ beats/min

2. Drag the vagus nerve stimulation electrode to the electrode holder to the right of the heart. Note that, when the electrode locks in place, the vagus nerve is draped over the electrode. Stimuli will go directly to the vagus nerve and indirectly to the heart.

3. Enter the number of ventricular contractions per minute (from the heart rate display) in the field below and then and click **Submit** to record your answer in the lab report.

_____ beats/min

> **? PREDICT Question 1**
> What do you think will happen if you apply multiple stimuli to the heart by indirectly stimulating the vagus nerve?

4. Click **Multiple Stimuli** to deliver electrical shocks to the vagus nerve at a rate of 50 stimuli/sec. The **Multiple Stimuli** button changes to a **Stop Stimuli** button as soon as it is clicked. Observe the effects of stimulation on the contractile activity and, after waiting at least 20 seconds (the tracing will make two full sweeps across the oscilloscope), click **Stop Stimuli** to stop the stimuli.

After you complete the experiment, take the online **Post-lab Quiz** for Activity 2.

Activity Questions

1. Describe how stimulation of the vagus nerves affects the heart rate.

_____

_____

2. How does the sympathetic nervous system affect heart rate and the force of contraction?

_____

_____

3. Describe the mechanism of vagal escape.

_____

_____

4. What would happen to the heart rate if the vagus nerve were cut?

_____

_____

_____

ACTIVITY 3

Examining the Effect of Temperature on Heart Rate

OBJECTIVES

1. To define the terms *hyperthermia* and *hypothermia*.
2. To contrast the terms *homeothermic* and *poikilothermic*.
3. To understand the effect that temperature has on the frog heart.
4. To understand the effect that temperature could have on the human heart.

Introduction

Humans are **homeothermic**, which means that the human body maintains an internal body temperature within the 35.8–38.2°C range even though the external temperature is changing. When the external temperature is elevated, the hypothalamus is signaled to activate heat-releasing mechanisms, such as sweating and vasodilation, to maintain the body's internal temperature. During extreme external temperature conditions, the body might not be able to maintain homeostasis and either **hyperthermia** (elevated body temperature) or **hypothermia** (low body temperature) could result. In contrast, the frog is a **poikilothermic** animal. Its internal body temperature changes depending on the temperature of its external environment because it lacks internal homeostatic regulatory mechanisms.

Ringer's solution, also known as Ringer's irrigation, consists of essential electrolytes (chloride, sodium, potassium, calcium, and magnesium) in a physiological solution and is required to keep the isolated, intact heart viable. In this activity you will explore the effect of temperature on heart rate using a Ringer's solution incubated at different temperatures.

EQUIPMENT USED The following equipment will be depicted on-screen: oscilloscope display—displays the contractile activity from the frog heart; electrical stimulator—used to a apply electrical shocks to the frog heart; electrode holder—locks electrodes in place for stimulation; external stimulation electrode; apparatus for sustaining an isolated, intact frog heart—includes 5°C, 23°C, and 32°C Ringer's solution; frog heart.

Experiment Instructions

Go to the home page in the PhysioEx software and click **Exercise 6: Cardiovascular Physiology.** Click **Activity 3: Examining the Effect of Temperature on Heart Rate,** and take the online **Pre-lab Quiz** for Activity 3.

After you take the online Pre-lab Quiz, click the **Experiment** tab and begin the experiment. The experiment instructions are reprinted here for your reference. The opening screen for the experiment is shown below.

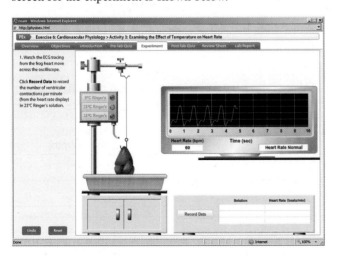

1. Watch the contractile activity from the frog heart on the oscilloscope. Click **Record Data** to record the number of ventricular contractions per minute (from the heart rate display) in 23°C Ringer's solution.

? PREDICT Question 1
What effect will decreasing the temperature of the Ringer's solution have on the heart rate of the frog?

---

2. Click **5°C Ringer's** to observe the effects of lowering the temperature.

3. When the heart activity display reads *Heart Rate Stable,* click **Record Data** to display your results in the grid (and record your results in Chart 3).

CHART 3	Effect of Temperature on Heart Rate
Solution	Heart rate (beats/min)

4. Click **23°C Ringer's** to bathe the heart and return it to room temperature. When the heart activity display reads *Heart Rate Normal,* you can proceed.

? PREDICT Question 2
What effect will increasing the temperature of the Ringer's solution have on the heart rate of the frog?

---

5. Click **32°C Ringer's** to observe the effects of increasing the temperature.

6. When the heart activity display reads *Heart Rate Stable,* click **Record Data** to display your results in the grid (and record your results in Chart 3).

After you complete the experiment, take the online **Post-lab Quiz** for Activity 3.

Activity Questions

1. Explain the importance of Ringer's solution (essential electrolytes in physiological saline) in maintaining the autorhythmicity of the heart.

2. Describe the effect of lower temperature on heart rate.

_____

_____

3. Explain the effect that fever would have on heart rate. Explain why.

_____

_____ ▬

Examining the Effects of Chemical Modifiers on Heart Rate

OBJECTIVES

1. To distinguish between cholinergic and adrenergic modifiers of heart rate.
2. To define agonist and antagonist modifiers of heart rate.
3. To observe the effects of epinephrine, pilocarpine, atropine, and digitalis on heart rate.
4. To relate chemical modifiers of the heart rate to sympathetic and parasympathetic activation.

Introduction

Although the heart does not need external stimulation to beat, it can be affected by extrinsic controls, most notably, the autonomic nervous system. The sympathetic nervous system is activated in times of "fight or flight," and sympathetic nerve fibers release **norepinephrine** (also known as **noradrenaline**) and **epinephrine** (also known as **adrenaline**) at their cardiac synapses.

Norepinephrine and epinephrine increase the frequency of action potentials by binding to β_1 adrenergic receptors embedded in the plasma membrane of **sinoatrial (SA) node** (pacemaker) cells. Working through a cAMP second-messenger mechanism, binding of the ligand opens sodium and calcium channels, increasing the rate of depolarization and shortening the period of repolarization, thus increasing the heart rate.

The parasympathetic nervous system, our "resting and digesting branch," usually dominates, and parasympathetic nerve fibers release **acetylcholine** at their cardiac synapses. Acetylcholine decreases the frequency of action potentials by binding to muscarinic cholinergic receptors embedded in the plasma membrane of the SA node cells. Acetylcholine indirectly opens potassium channels and closes calcium and sodium channels, decreasing the rate of depolarization and, thus, decreasing heart rate.

Chemical modifiers that inhibit, mimic, or enhance the action of acetylcholine in the body are labeled **cholinergic**. Chemical modifiers that inhibit, mimic, or enhance the action of epinephrine in the body are **adrenergic**. If the modifier works in the same fashion as the neurotransmitter (acetylcholine or norepinephrine), it is an **agonist**. If the modifier works in opposition to the neurotransmitter, it is an **antagonist**. In this activity you will explore the effects of pilocarpine, atropine, epinephrine, and digitalis on heart rate.

EQUIPMENT USED The following equipment will be depicted on-screen: oscilloscope display—displays the contractile activity; apparatus for sustaining an isolated intact frog heart—includes 23°C Ringer's solution; pilocarpine; atropine; epinephrine; digitalis; frog heart.

Experiment Instructions

Go to the home page in the PhysioEx software and click **Exercise 6: Cardiovascular Physiology.** Click **Activity 4: Examining the Effects of Chemical Modifiers on Heart Rate,** and take the online **Pre-lab Quiz** for Activity 4.

After you take the online Pre-lab Quiz, click the **Experiment** tab and begin the experiment. The experiment instructions are reprinted here for your reference. The opening screen for the experiment is shown below.

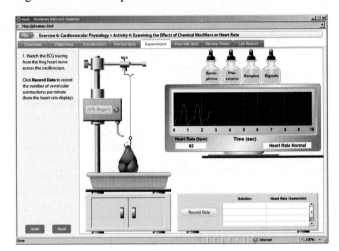

1. Watch the contractile activity from the frog heart on the oscilloscope. Click **Record Data** to record the number of ventricular contractions per minute (from the heart rate display) and record your results in Chart 4.

2. Drag the dropper cap of the epinephrine bottle to the frog heart to release epinephrine onto the heart.

3. Observe the contractile activity and the heart activity display. When the heart activity display reads *Heart Rate Stable*, click **Record Data** to display your results in the grid (and record your results in Chart 4).

CHART 4	Effects of Chemical Modifiers on Heart Rate
Solution	Heart rate (beats/min)

4. Click **23°C Ringer's** (room temperature) to bathe the heart and flush out the epinephrine. When the heart activity display reads *Heart Rate Normal*, you can proceed.

> **? PREDICT Question 1**
> Pilocarpine is a cholinergic drug, an acetylcholine agonist. Predict the effect that pilocarpine will have on heart rate.
>
> _____

5. Drag the dropper cap of the pilocarpine bottle to the frog heart to release pilocarpine onto the heart.

6. Observe the contractile activity and the heart activity display. When the heart activity display reads *Heart Rate Stable*, click **Record Data** to display your results in the grid (and record your results in Chart 4).

7. Click **23°C Ringer's** (room temperature) to bathe the heart and flush out the pilocarpine. When the heart activity display reads *Heart Rate Normal*, you can proceed.

> **? PREDICT Question 2**
> Atropine is another cholinergic drug, an acetylcholine antagonist. Predict the effect that atropine will have on heart rate.
>
> _____

8. Drag the dropper cap of the atropine bottle to the frog heart to release atropine onto the heart.

9. Observe the contractile activity and the heart activity display. When the heart activity display reads *Heart Rate Stable*, click **Record Data** to display your results in the grid (and record your results in Chart 4).

10. Click **23°C Ringer's** (room temperature) to bathe the heart and flush out the atropine. When the heart activity display reads *Heart Rate Normal*, you can proceed.

11. Drag the dropper cap of the digitalis bottle to the frog heart to release digitalis onto the heart.

12. Observe the contractile activity and the heart activity display. When the heart activity display reads *Heart Rate Stable*, click **Record Data** to display your results in the grid (and record your results in Chart 4).

After you complete the experiment, take the online **Post-lab Quiz** for Activity 4.

Activity Questions

1. Define *agonist* and *antagonist*. Clearly distinguish between the two and give examples used in this activity.

_____

_____

2. Describe the effect of epinephrine on heart rate and force of contraction.

_____

_____

3. What is the effect of atropine on heart rate?

_____

_____

4. Describe the effect of digitalis on heart rate and force of contraction.

_____

_____ ▬

A C T I V I T Y   5

Examining the Effects of Various Ions on Heart Rate

OBJECTIVES

1. To understand the movement of ions that occurs during the cardiac action potential.
2. To describe the potential effect of potassium, sodium, and calcium ions on heart rate.
3. To explain how calcium channel blockers might be used pharmaceutically to treat heart patients.
4. To define the terms *inotropic* and *chronotropic*.

Introduction

In cardiac muscle cells, action potentials are caused by changes in permeability to ions due to the opening and closing of ion channels. The permeability changes that occur for the cardiac muscle cell involve potassium, sodium, and calcium ions. The concentration of potassium is greater inside the cardiac muscle cell than outside the cell. Sodium and calcium are present in larger quantities outside the cell than inside the cell.

The resting cell membrane favors the movement of potassium more than sodium or calcium. Therefore, the resting membrane potential of cardiac cells is determined mainly by the ratio of extracellular and intracellular concentrations of potassium. View Table 6.1 for a summary of the phases of the cardiac action potential and ion movement during each phase.

Calcium channel blockers are used to treat high blood pressure and abnormal heart rates. They block the movement of calcium through its channels throughout all phases of the cardiac action potentials. Consequently, because less calcium gets through, both the rate of depolarization and the force of the contraction are reduced. Modifiers that affect heart rate are **chronotropic**, and modifiers that affect the force of contraction are **inotropic**. Modifiers that lower heart rate are negative chronotropic, and modifiers that increase heart rate are positive chronotropic. The same adjectives describe inotropic modifiers. Therefore, negative inotropic drugs decrease the force of contraction of the heart and positive inotropic drugs increase the force of contraction of the heart.

TABLE 6.1

Phase of cardiac action potential	Ion movement
Phase 0 (rapid depolarization)	Sodium moves in
Phase 1 (small repolarization)	Sodium movement decreases
Phase 2 (plateau)	Potassium movement out decreases Calcium moves in
Phase 3 (repolarization)	Potassium moves out Calcium movement decreases
Phase 4 (resting potential)	Potassium moves out Little sodium or calcium moves in

EQUIPMENT USED The following equipment will be depicted on-screen: oscilloscope display—displays the contractile activity from the frog heart; apparatus for sustaining frog heart—includes 23°C Ringer's solution; calcium ions; sodium ions; potassium ions; frog heart.

Experiment Instructions

Go to the home page in the PhysioEx software and click **Exercise 6: Cardiovascular Physiology.** Click **Activity 5: Examining the Effects of Various Ions on Heart Rate,** and take the online **Pre-lab Quiz** for Activity 5.

After you take the online Pre-lab Quiz, click the **Experiment** tab and begin the experiment. The experiment instructions are reprinted here for your reference. The opening screen for the experiment is shown below.

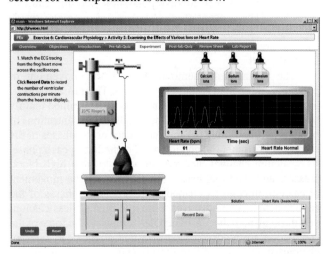

1. Watch the contractile activity from the frog heart on the oscilloscope. Click **Record Data** to record the number of ventricular contractions per minute (from the heart rate display).

> **? PREDICT Question 1**
> Because calcium channel blockers are negative chronotropic and negative inotropic, what effect do you think increasing the concentration of calcium will have on heart rate?
> _____

2. Drag the dropper cap of the calcium ions bottle to the frog heart to release calcium ions onto the heart. Note the change in heart rate after you drop the calcium ions onto the heart.

3. When the heart activity display reads *Heart Rate Stable,* click **Record Data** to display your results in the grid (and record your results in Chart 5).

CHART 5	Effects of Various Ions on Heart Rate	
Solution		Heart rate (beats/min)

4. **Click 23°C Ringer's** (room temperature) to bathe the heart and flush out the calcium. When the heart activity display reads *Heart Rate Normal,* you can proceed.

5. Drag the dropper cap of the sodium ions bottle to the frog heart to release sodium ions onto the heart. Note the immediate change in the heart rate and the change in heart rate over time after you drop the sodium ions onto the heart.

6. After waiting at least 20 seconds (the tracing will make two full sweeps across the oscilloscope), click **Record Data** to display your results in the grid (and record your results in Chart 5).

7. Click **23°C Ringer's** (room temperature) to bathe the heart and flush out the sodium. When the heart activity display reads *Heart Rate Normal,* you can proceed.

> **? PREDICT Question 2**
> Excess potassium outside of the cardiac cell decreases the resting potential of the plasma membrane, thus decreasing the force of contraction. What effect (if any) do you think it will *initially* have on heart rate?
> _____

8. Drag the dropper cap of the potassium ions bottle to the frog heart to release potassium ions onto the heart. Note the immediate change in heart rate and the change in heart rate over time after you drop the potassium ions onto the heart.

9. After waiting at least 20 seconds (the tracing will make two full sweeps across the oscilloscope), click **Record Data** to display your results in the grid (and record your results in Chart 5).

After you complete the experiment, take the online **Post-lab Quiz** for Activity 5.

Activity Questions

1. Define chronotropic and inotropic effects on the heart.

_____

_____

2. Describe the effect of adding calcium ions to the frog heart.

_____

_____

3. Calcium channel blockers are often used to treat high blood pressure. Explain how their effects would benefit individuals with high blood pressure.

_____

_____

4. Describe the initial effect of adding potassium ions to the frog heart.

_____

_____

Cardiovascular Physiology

NAME_____

LAB TIME/DATE _____

ACTIVITY 1   Investigating the Refractory Period of Cardiac Muscle

1. Explain why the larger waves seen on the oscilloscope represent ventricular contraction.

   _____

   _____

2. Explain why the amplitude of the wave did not change when you increased the frequency of the stimulation. (Hint: Relate your

   response to the refractory period of the cardiac action potential.) How well did the results compare with your prediction?

   _____

   _____

3. Why is it only possible to induce an extrasystole during relaxation? _____

   _____

4. Explain why wave summation and tetanus are not possible in cardiac muscle tissue. How well did the results compare with

   your prediction? _____

   _____

ACTIVITY 2   Examining the Effect of Vagus Nerve Stimulation

1. Explain the effect that extreme vagus nerve stimulation had on the heart. How well did the results compare with your prediction?

   _____

   _____

2. Explain two ways that the heart can overcome excessive vagal stimulation. _____

   _____

3. Describe how the sympathetic and parasympathetic nervous systems work together to regulate heart rate. _____

   _____

4. What do you think would happen to the heart rate if the vagus nerve was cut? _____

   _____

A C T I V I T Y 3   Examining the Effect of Temperature on Heart Rate

1. Explain the effect that decreasing the temperature had on the frog heart. How do you think the human heart would respond?

   How well did the results compare with your prediction? _____

   _____

2. Describe why Ringer's solution is required to maintain heart contractions. _____

   _____

3. Explain the effect that increasing the temperature had on the frog heart. How do you think the human heart would respond?

   How well did the results compare with your prediction? _____

   _____

A C T I V I T Y 4   Examining the Effects of Chemical Modifiers on Heart Rate

1. Describe the effect that pilocarpine had on the heart and why it had this effect. How well did the results compare with your

   prediction? _____

   _____

2. Atropine is an acetylcholine antagonist. Does atropine inhibit or enhance the effects of acetylcholine? Describe your results

   and how they correlate with how the drug works. How well did the results compare with your prediction? _____

   _____

3. Describe the benefits of administering digitalis. _____

   _____

4. Distinguish between cholinergic and adrenergic chemical modifiers. Include examples of each in your discussion. _____

   _____

A C T I V I T Y 5   Examining the Effects of Various Ions on Heart Rate

1. Describe the effect that increasing the calcium ions had on the heart. How well did the results compare with your prediction?

   _____

2. Describe the effect that increasing the potassium ions initially had on the heart in this activity. Relate this to the resting

   membrane potential of the cardiac muscle cell. How well did the results compare with your prediction? _____

   _____

3. Describe how calcium channel blockers are used to treat patients and why. _____

   _____

Respiratory System Mechanics

Exercise Overview

The physiological function of the respiratory system is essential to life. If problems develop in most other physiological systems, we can survive for some time without addressing them. But if a persistent problem develops within the respiratory system (or the circulatory system), death can occur in minutes.

The primary role of the respiratory system is to distribute oxygen to, and remove carbon dioxide from, *all* the cells of the body. The respiratory system works together with the circulatory system to achieve this. **Respiration** includes **ventilation,** or the movement of air into and out of the lungs (breathing), and the transport (via blood) of oxygen and carbon dioxide between the lungs and body cells (view Figure 7.1). The heart pumps deoxygenated blood to pulmonary capillaries, where gas exchange occurs between blood and **alveoli** (air sacs in the lungs), thus oxygenating the blood. The heart then pumps the oxygenated blood to body tissues, where oxygen is used for cell metabolism. At the same time, carbon dioxide (a waste product of metabolism) from body tissues diffuses into the blood. This carbon dioxide–enriched, oxygen-reduced blood then returns to the heart, completing the circuit.

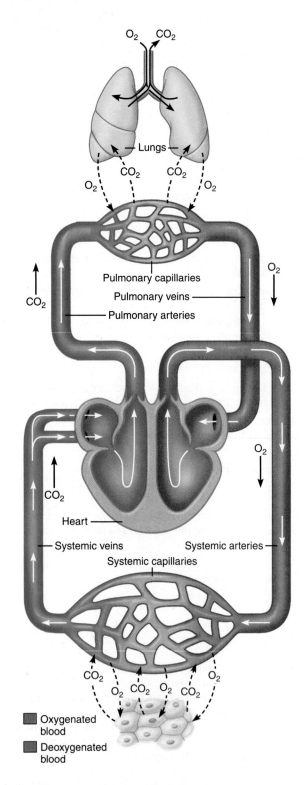

FIGURE 7.1 **Relationship between external respiration and internal respiration.**

Ventilation is the result of skeletal muscle contraction. When the **diaphragm**—a dome-shaped muscle that divides the thoracic and abdominal cavities—and the **external intercostal muscles** contract, the volume in the thoracic cavity increases. This increase in thoracic volume reduces the pressure in the thoracic cavity, allowing atmospheric gas to enter the lungs

(a process called **inspiration**). When the diaphragm and the external intercostals relax, the pressure in the thoracic cavity increases as the volume decreases, forcing air out of the lungs (a process called **expiration**). Inspiration is considered an *active* process because muscle contraction requires the use of ATP, whereas expiration is usually considered a *passive* process because the muscles relax, rather than contract. When a person is running, however, expiration becomes an active process, resulting from the contraction of **internal intercostal muscles** and **abdominal muscles.** In this case, both inspiration and expiration are considered *active* processes because muscle contraction is needed for both.

The amount of air that flows into and out of the lungs in 1 minute is the pulmonary **minute ventilation,** which is calculated by multiplying the **frequency of breathing** by the volume of each breath (the **tidal volume**). Ventilation must be regulated at all times to maintain oxygen in arterial blood and carbon dioxide in venous blood at their normal levels—that is, at their normal **partial pressures.** The *partial pressure* of a gas is the proportion of pressure that the gas exerts in a mixture. For example, in the atmosphere at sea level, the total pressure is 760 mm Hg. Oxygen makes up 21% of the total atmosphere and, therefore, has a partial pressure (P_{O_2}) of 160 mm Hg (760 mm Hg $\times$ 0.21).

Oxygen and carbon dioxide diffuse down their partial pressure gradients, from high partial pressures to low partial pressures. Oxygen diffuses from the alveoli of the lungs into the blood, where it can dissolve in plasma and attach to hemoglobin, and then diffuses from the blood into the tissues. Carbon dioxide (produced by the metabolic reactions of the tissues) diffuses from the tissues into the blood and then diffuses from the blood into the alveoli for export from the body.

In this exercise you will investigate the basic mechanics and regulation of the respiratory system. The concepts you will explore with a simulated lung will help you understand the operation of the human respiratory system in better detail.

ACTIVITY 1

Measuring Respiratory Volumes and Calculating Capacities

OBJECTIVES

1. To understand the use of the terms *ventilation, inspiration, expiration, diaphragm, external intercostals, internal intercostals, abdominal-wall muscles, expiratory reserve volume (ERV), forced vital capacity (FVC), tidal volume (TV), inspiratory reserve volume (IRV), residual volume (RV),* and *forced expiratory volume in one second (FEV$_1$).*

2. To understand the roles of skeletal muscles in the mechanics of breathing.

3. To understand the volume and pressure changes in the thoracic cavity during ventilation of the lungs.

4. To understand the effects of airway radius and, thus, resistance on airflow.

Introduction

The two phases of **ventilation,** or breathing, are (1) **inspiration,** during which air is taken into the lungs, and (2) **expiration,** during which air is expelled from the lungs. Inspiration occurs as the **external intercostal muscles** and the **diaphragm** contract. The diaphragm, normally a dome-shaped muscle, flattens as it moves inferiorly while the external intercostal muscles, situated between the ribs, lift the rib cage. These cooperative actions increase the thoracic volume. Air rushes into the lungs because this increase in thoracic volume creates a partial vacuum.

During quiet expiration, the inspiratory muscles relax, causing the diaphragm to rise superiorly and the chest wall to move inward. Thus, the **thorax** returns to its normal shape because of the elastic properties of the lung and thoracic wall. As in a deflating balloon, the pressure in the lungs rises, forcing air out of the lungs and airways. Although expiration is normally a *passive* process, **abdominal-wall muscles** and the **internal intercostal muscles** can also contract during expiration to force additional air from the lungs. Such forced expiration occurs, for example, when you exercise, blow up a balloon, cough, or sneeze.

Normal, quiet breathing moves about 500 ml (0.5 liter) of air (the **tidal volume**) into and out of the lungs with each breath, but this amount can vary due to a person's size, sex, age, physical condition, and immediate respiratory needs. In this activity you will measure the following respiratory volumes (the values given for the normal adult male and female are approximate).

Tidal volume (TV): Amount of air inspired and then expired with each breath under resting conditions (500 ml)

Inspiratory reserve volume (IRV): Amount of air that can be forcefully inspired after a normal tidal volume inspiration (male, 3100 ml; female, 1900 ml)

Expiratory reserve volume (ERV): Amount of air that can be forcefully expired after a normal tidal volume expiration (male, 1200 ml; female, 700 ml)

Residual volume (RV): Amount of air remaining in the lungs after forceful and complete expiration (male, 1200 ml; female, 1100 ml)

Respiratory capacities are calculated from the respiratory volumes. In this activity you will calculate the following respiratory capacities.

Total lung capacity (TLC): Maximum amount of air contained in lungs after a maximum inspiratory effort: TLC = TV + IRV + ERV + RV (male, 6000 ml; female, 4200 ml)

Vital capacity (VC): Maximum amount of air that can be inspired and then expired with maximal effort: VC = TV + IRV + ERV (male, 4800 ml; female 3100 ml)

You will also perform two pulmonary function tests in this activity.

Forced vital capacity (FVC): Amount of air that can be expelled when the subject takes the deepest possible inspiration and forcefully expires as completely and rapidly as possible

Forced expiratory volume (FEV$_1$): Measures the percentage of the vital capacity that is expired during 1 second of the FVC test (normally 75%–85% of the vital capacity)

EQUIPMENT USED The following equipment will be depicted on-screen: simulated human lungs suspended in a glass bell jar; rubber diaphragm—used to seal the jar and change the volume and, thus, pressure in the jar (As the diaphragm moves inferiorly, the volume in the bell jar increases and the pressure drops slightly, creating a partial vacuum in the bell jar. This partial vacuum causes air to be sucked into the tube at the top of the bell jar and then into the simulated lungs. As the diaphragm moves up, the decreasing volume and rising pressure within the bell jar forces air out of the lungs.); adjustable airflow tube—connects the lungs to the atmosphere; oscilloscope; three different breathing patterns: normal tidal volumes, expiratory reserve volume (ERV), and forced vital capacity (FVC).

Experiment Instructions

Go to the home page in the PhysioEx software and click **Exercise 7: Respiratory System Mechanics.** Click **Activity 1: Measuring Respiratory Volumes and Calculating Capacities,** and take the online **Pre-lab Quiz** for Activity 1.

After you take the online Pre-lab Quiz, click the **Experiment** tab and begin the experiment. The experiment instructions are reprinted here for your reference. The opening screen for the experiment is shown below.

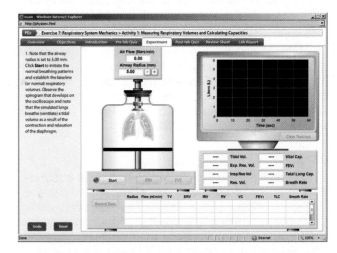

1. Note that the airway radius is set to 5.00 mm. Click **Start** to initiate the normal breathing patterns and establish the baseline (or normal) respiratory volumes. Observe the spirogram that develops on the oscilloscope and note that the simulated lungs breathe (ventilate) a tidal volume as a result of the contraction and relaxation of the diaphragm.

2. Click **Record Data** to display your results in the grid (and record your results in Chart 1).

3. Click **Clear Tracings** to clear the spirogram on the oscilloscope.

4. You will now complete the measurement of respiratory volumes and determine the respiratory capacities. First, click **Start** to initiate the normal breathing pattern. After 10 seconds, click **ERV.** Wait another 10 seconds and then click **FVC** to

CHART 1	Respiratory Volumes and Capacities							
Radius (mm)	Flow (ml/min)	TV (ml)	ERV (ml)	IRV (ml)	RV (ml)	VC (ml)	FEV$_1$ (ml)	TLC (ml)

complete the measurement of respiratory volumes. When you click ERV, the program will simulate forced expiration using the contraction of the internal intercostal muscles and abdominal-wall muscles. When you click FVC, the lungs will first inspire maximally and then expire fully to demonstrate forced vital capacity.

5. Note that, in addition to the tidal volume, the expiratory reserve volume, inspiratory reserve volume, and residual volume were measured. The vital capacity and total lung capacity were calculated from those volumes. Click **Record Data** to display your results in the grid (and record your results in Chart 1).

6. Minute ventilation is the amount of air that flows into and then out of the lungs in a minute. Minute ventilation (ml/min) = TV (ml/breath) × BPM (breaths/min). Using the values from the second recorded measurement, enter the minute ventilation in the field below and then click **Submit** to record your answer in the lab report. _____ ml/min

? PREDICT Question 1
Lung diseases are often classified as obstructive or restrictive. An **obstructive** disease affects *airflow*, and a **restrictive** disease usually reduces *volumes and capacities*. Although they are not diagnostic, pulmonary function tests such as forced expiratory volume (FEV$_1$) can help a clinician determine the difference between obstructive and restrictive diseases. Specifically, an FEV$_1$ is the forced volume expired in 1 second.
   In obstructive diseases such as chronic bronchitis and asthma, airway radius is decreased. Thus, FEV$_1$ will:

_____.

7. You will now explore what effect changing the airway radius has on pulmonary function. Decrease the airway radius to 4.50 mm by clicking the − button beneath the airway radius display.

8. Click **Start** to initiate the normal breathing pattern. After 10 seconds, click **ERV.** Wait another 10 seconds and then click **FVC.** The FEV$_1$ will appear in the FEV$_1$ display beneath the oscilloscope.

9. Click **Record Data** to display your results in the grid (and record your results in Chart 1).

10. You will now gradually decrease the airway radius.

   • Decrease the airway radius by 0.50 mm by clicking the − button beneath the airway radius display.

   • Click **Start** to initiate the normal breathing pattern. After 10 seconds, click **ERV.** Wait another 10 seconds and then click **FVC.** The FEV$_1$ will appear in the FEV$_1$ display beneath the oscilloscope.

   • Click **Record Data** to display your results in the grid (and record your results in Chart 1).

   Repeat this step until you reach an airway radius of 3.00 mm.

11. A useful way to express FEV$_1$ is as a percentage of the forced vital capacity (FVC). Using the FEV$_1$ and FVC values from the data grid, calculate the FEV$_1$ (%) by dividing the FEV$_1$ volume by the FVC volume (in this case, the VC is equal to the FVC) and multiply by 100%. Enter the FEV$_1$ (%) for an airway radius of 5.0 mm in the field below and then click **Submit** to record your answer in the lab report.

FEV$_1$ (%) for an airway radius of 5.0 (mm): _____

12. Enter the FEV$_1$ (%) for an airway radius of 3.00 mm in the field below and then click **Submit** to record your answer in the lab report.

FEV$_1$ (%) for an airway radius of 3.00 (mm): _____

After you complete the experiment, take the online **Post-lab Quiz** for Activity 1.

Activity Questions

1. When you forcefully exhale your entire expiratory reserve volume, any air remaining in your lungs is called the residual volume (RV). Why is it impossible to further exhale the RV (that is, *where* is this air volume trapped, and *why* is it trapped)?

_____

_____

2. How do you measure a person's RV in a laboratory?

_____

_____

3. Draw a spirogram that depicts a person's volumes and capacities before and during a significant cough.

Comparative Spirometry

OBJECTIVES

1. To understand the terms _spirometry, spirogram, emphysema, asthma, inhaler, moderate exercise, heavy exercise, tidal volume (TV), expiratory reserve volume (ERV), inspiratory reserve volume (IRV), residual volume (RV), vital capacity (VC), total lung capacity (TLC), forced vital capacity (FVC),_ and _forced expiratory volume in one second (FEV$_1$)._

2. To observe and compare spirograms collected from resting, healthy patients to those taken from an emphysema patient.

3. To observe and compare spirograms collected from resting, healthy patients to those taken from a patient suffering an acute asthma attack.

4. To observe and compare the spirogram collected from an asthmatic patient _while_ suffering an acute asthma attack to that taken after the patient uses an inhaler for relief.

5. To observe and compare spirograms collected from volunteers engaged in moderate exercise and heavy exercise.

Introduction

In this activity you will explore the changes to normal respiratory volumes and capacities when pathophysiology develops and during aerobic exercise by recruiting volunteers to breathe into a water-filled spirometer. The spirometer is a device that measures the volume of air inspired and expired by the lungs over a specified period of time. Several lung capacities and flow rates can be calculated from this data to assess pulmonary function. With your knowledge of respiratory mechanics, you can predict, document, and explain changes to the volumes and capacities in each state.

Emphysema breathing: With emphysema, there is a significant loss of elastic recoil in the lung tissue. This loss of elastic recoil occurs as the disease destroys the walls of the alveoli. Airway resistance is also increased as the lung tissue in general becomes more flimsy and exerts less anchoring on the surrounding airways. Thus, the lung becomes overly compliant and expands easily. Conversely, a great effort is required to expire because the lungs can no longer passively recoil and deflate. Each expiration requires a noticeable and exhausting muscular effort, and a person with emphysema expires slowly.

Acute asthma attack breathing: During an acute asthma attack, bronchiole smooth muscle spasms and, thus, the airways become constricted (that is, reduced in diameter). They also become clogged with thick mucus secretions. These changes lead to significantly increased airway resistance.

Underlying these symptoms is an airway inflammatory response brought on by triggers such as allergens (for example, dust and pollen), extreme temperature changes, and even exercise. Like with emphysema, the airways collapse and pinch closed before a forced expiration is completed. Thus, the volumes and peak flow rates are significantly reduced during an asthma attack. Unlike with emphysema, the elastic recoil is not diminished in an acute asthma attack.

When an acute asthma attack occurs, many people seek to relieve symptoms with an inhaler, which atomizes the medication and allows for direct application onto the afflicted airways. Usually, the medication includes a smooth muscle relaxant (for example, a β_2 agonist or an acetylcholine antagonist) that relieves the bronchospasms and induces bronchiole dilation. The medication can also contain an anti-inflammatory agent, such as a corticosteroid, that suppresses the inflammatory response. The use of the inhaler reduces airway resistance.

Breathing during exercise: During _moderate_ aerobic exercise, the human body has an increased metabolic demand, which is met, in part, by changes in respiration. Specifically, both the rate of breathing and the tidal volume increase. These two respiratory variables do not increase by the same amount. The increase in the tidal volume is greater than the increase in the rate of breathing. During _heavy_ exercise, further changes in respiration are required to meet the extreme metabolic demands of the body. In this case both the rate of breathing and the tidal volume increase to their maximum tolerable limits.

> **EQUIPMENT USED** The following equipment will be depicted on-screen: a classic water-filled spirometer with an attached rotating drum that records the analog spirogram in real time; breathing patterns from a variety of patients: unforced breathing and forced vital capacity for a "normal" patient, a patient with emphysema, and a patient with asthma (during an attack and after using an inhaler); and the breathing patterns from a patient during moderate and heavy exercise.

Experiment Instructions

Go to the home page in the PhysioEx software and click **Exercise 7: Respiratory System Mechanics.** Click **Activity 2: Comparative Spirometry,** and take the online **Pre-lab Quiz** for Activity 2.

After you take the online Pre-lab Quiz, click the **Experiment** tab and begin the experiment. The experiment instructions are reprinted here for your reference. The opening screen for the experiment is shown below.

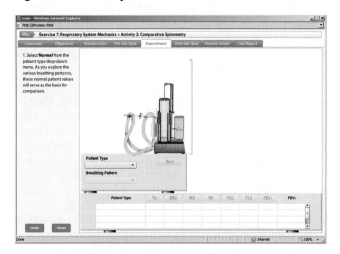

1. Select **Normal** from the patient type drop-down menu. As you explore the various breathing patterns, these normal patient values will serve as the basis for comparison.

2. Select **Unforced Breathing** from the breathing pattern drop-down menu.

3. Click **Start** to record the patient's unforced breathing pattern and watch as the drum starts turning and the spirogram develops on the paper rolling off the drum.

4. Note the volume levels (in milliliters) on the Y-axis of the spirogram. When half the screen is filled with unforced tidal volumes and the spirogram has paused, select **Forced Vital Capacity** from the breathing pattern drop-down menu.

5. Click **Start** to record the patient's forced vital capacity. The spirogram ends as the paper rolls to the right edge of the screen.

6. Click on each of the buttons in the data recorder to measure respiratory volumes and capacities. Start with tidal volume (TV) and work your way to the right. When you measure each volume or capacity, (1) a bracket appears on the

spirogram to indicate where that measurement originates and (2) the value (in milliliters) displays in the grid. After you complete all the measurements, the FEV_1 (%) ratio will automatically be calculated. The FEV_1 (%) = (FEV_1/FVC) $\times$ 100%. Record your results in Chart 2.

> **? PREDICT Question 1**
> With emphysema, there is a significant loss of elastic recoil in the lung tissue and a noticeable, exhausting muscular effort is required for each expiration. Inspiration actually becomes easier because the lung is now overly compliant. What lung values will change (from those of the normal patient) in the spirogram when the patient with emphysema is selected?
> _____

7. Select **Emphysema** from the patient type drop-down menu.

8. Select **Unforced Breathing** from the breathing pattern drop-down menu.

9. Click **Start** to record the patient's unforced breathing pattern and watch as the drum starts turning and the spirogram develops on the paper rolling off the drum.

10. Note the volume levels on the Y-axis of the spirogram. When half the screen is filled with unforced tidal volumes and the spirogram has paused, select **Forced Vital Capacity** from the breathing pattern drop-down menu.

11. Click **Start** to record the patient's forced vital capacity. The spirogram ends as the paper rolls to the right edge of the screen.

12. Click on each of the buttons in the data recorder to measure respiratory volumes and capacities. Start with tidal volume (TV) and work your way to the right. Record your results in Chart 2.

CHART 2	Spirometry Results							
Patient type	TV (ml)	ERV (ml)	IRV (ml)	RV (ml)	FVC (ml)	TLC (ml)	FEV_1 (ml)	FEV_1 (%)
Normal								
Emphysema								
Acute asthma attack								
Plus inhaler								
Moderate exercise								
Heavy exercise								

? PREDICT **Question 2**
During an acute asthma attack, airway resistance is significantly increased by (1) increased thick mucous secretions and (2) airway smooth muscle spasms. What lung values will change (from those of the normal patient) in the spirogram for a patient suffering an acute asthma attack?

13. Select **Acute Asthma Attack** from the patient type drop-down menu.

14. Select **Unforced Breathing** from the breathing pattern drop-down menu.

15. Click **Start** to record the patient's uforced breathing pattern and watch as the drum starts turning and the spirogram develops on the paper rolling off the drum.

16. Note the volume levels on the Y-axis of the spirogram. When half the screen is filled with unforced tidal volumes and the spirogram has paused, select **Forced Vital Capacity** from the breathing pattern drop-down menu.

17. Click **Start** to record the patient's forced vital capacity. The spirogram ends as the paper rolls to the right edge of the screen.

18. Click on each of the buttons in the data recorder to measure respiratory volumes and capacities. Start with tidal volume (TV) and work your way to the right. Record your results in Chart 2.

? PREDICT **Question 3**
When an acute asthma attack occurs, many people seek relief from the increased airway resistance by using an inhaler. This device atomizes the medication and induces bronchiole dilation (though it can also contain an anti-inflammatory agent). What lung values will change *back* to those of the normal patient in the spirogram after the asthma patient uses an inhaler?

19. Select **Plus Inhaler** from the patient type drop-down menu.

20. Select **Unforced Breathing** from the breathing pattern drop-down menu.

21. Click **Start** to record the patient's unforced breathing pattern and watch as the drum starts turning and the spirogram develops on the paper rolling off the drum.

22. Note the volume levels on the Y-axis of the spirogram. When half the screen is filled with unforced tidal volumes and the spirogram has paused, select **Forced Vital Capacity** from the breathing pattern drop-down menu.

23. Click **Start** to record the patient's forced vital capacity. The spirogram ends as the paper rolls to the right edge of the screen.

24. Click on each of the buttons in the data recorder to measure respiratory volumes and capacities. Start with tidal volume (TV) and work your way to the right. Record your results in Chart 2.

? PREDICT **Question 4**
During moderate aerobic exercise, the human body will change its respiratory cycle in order to meet increased metabolic demands. During heavy exercise, further changes in respiration are required to meet the extreme metabolic demands of the body. Which lung value will change more during moderate exercise, the ERV or the IRV?

25. Select **Moderate Exercise** from the patient type drop-down menu. Note that the selection of a breathing pattern is not applicable because our central nervous system automatically adjusts and maintains the depth and frequency of breathing to meet the increased metabolic demands while we exercise. We do not normally alter this pattern with conscious intervention.

26. Click **Start** to record the patient's breathing pattern and watch as the drum starts turning and the spirogram develops on the paper rolling off the drum.

27. Click on each of the buttons in the data recorder to measure respiratory volumes and capacities. Start with tidal volume (TV) and work your way to the right. *ND* indicates this measurement or calculation was not done. Record your results in Chart 2.

28. Select **Heavy Exercise** from the patient type drop-down menu.

29. Click **Start** to record the patient's breathing pattern and watch as the drum starts turning and the spirogram develops on the paper rolling off the drum.

30. Click on each of the buttons in the data recorder to measure respiratory volumes and capacities. Start with tidal volume (TV) and work your way to the right. Record your results in Chart 2.

After you complete the experiment, take the online **Post-lab Quiz** for Activity 2.

Activity Questions

1. Why is residual volume (RV) above normal in a patient with emphysema?

2. Why did the asthmatic patient's inhaler medication fail to return all volumes and capacities to normal values right away?

3. Looking at the spirograms generated in this activity, state an easy way to determine whether a person's exercising effort is moderate or heavy.

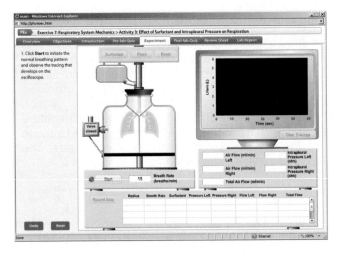

ACTIVITY 3

Effect of Surfactant and Intrapleural Pressure on Respiration

OBJECTIVES

1. To understand the terms *surfactant, surface tension, intrapleural space, intrapleural pressure, pneumothorax,* and *atelectasis.*

2. To understand the effect of surfactant on surface tension and lung function.

3. To understand how negative intrapleural pressure prevents lung collapse.

Introduction

At any gas-liquid boundary, the molecules of the liquid are attracted more strongly to each other than they are to the gas molecules. This unequal attraction produces tension at the liquid surface, called **surface tension.** Because surface tension resists any force that tends to increase surface area of the gas-liquid boundary, it acts to decrease the size of hollow spaces, such as the alveoli, or microscopic air spaces within the lungs.

If the film lining the air spaces in the lung were pure water, it would be very difficult, if not impossible, to inflate the lungs. However, the aqueous film covering the alveolar surfaces contains **surfactant,** a detergent-like mixture of lipids and proteins that decreases surface tension by reducing the attraction of water molecules to each other. You will explore the importance of surfactant in this activity.

Between breaths, the pressure in the pleural cavity, the **intrapleural pressure,** is less than the pressure in the alveoli. Two forces cause this negative pressure condition: (1) the tendency of the lung to recoil because of its elastic properties and the surface tension of the alveolar fluid and (2) the tendency of the compressed chest wall to recoil and expand outward. These two forces pull the lungs away from the thoracic wall, creating a partial vacuum in the pleural cavity.

Because the pressure in the intrapleural space is lower than atmospheric pressure, any opening created in the pleural membranes equalizes the intrapleural pressure with atmospheric pressure by allowing air to enter the pleural cavity, a condition called **pneumothorax.** A pneumothorax can then lead to lung collapse, a condition called **atelectasis.** In this activity, the **intrapleural space** is the space between the wall of the glass bell jar and the outer wall of the lung it contains.

EQUIPMENT USED The following equipment will be depicted on-screen: simulated human lungs suspended in a glass bell jar; rubber diaphragm—used to seal the jar and change the volume and, thus, pressure in the jar (As the diaphragm moves inferiorly, the volume in the bell jar increases and the pressure drops slightly, creating a partial vacuum in the bell jar. This partial vacuum causes air to be sucked into the tube at the top of the bell jar and then into the simulated lungs. As the diaphragm moves up, the decreasing volume and rising pressure within the bell jar forces air out of the lungs.); valve—allows intrapleural pressure in the left side of bell jar to equalize with atmospheric pressure; surfactant—amphipathic lipids (dipalmitoylphosphatidylcholine, phosphatidylglycerol, and palmitic acid) and short, synthetic peptides in a mixture that mimics the surfactant found in human lungs (surfactant molecules reduce surface tension in alveoli by adsorbing to the air-water interface, with their hydrophilic parts in the water and their hydrophobic parts facing toward the air); oscilloscope.

Experiment Instructions

Go to the home page in the PhysioEx software and click **Exercise 7: Respiratory System Mechanics.** Click **Activity 3: Effect of Surfactant and Intrapleural Pressure on Respiration,** and take the online **Pre-lab Quiz** for Activity 3.

After you take the online Pre-lab Quiz, click the **Experiment** tab and begin the experiment. The experiment instructions are reprinted here for your reference. The opening screen for the experiment is shown below.

1. Click **Start** to initiate the normal breathing pattern and observe the tracing that develops on the oscilloscope.

2. Click **Record Data** to display your results in the grid (and record your results in Chart 3). This data represents breathing in the absence of surfactant.

3. Click **Surfactant** twice to dispense two aliquots of the synthetic lipids and peptides onto the interior lining of the lungs.

4. Click **Start** to initiate breathing in the presence of surfactant and observe the tracing that develops.

5. Click **Record Data** to display your results in the grid (and record your results in Chart 3).

CHART 3	Effect of Surfactant and Intrapleural Pressure on Respiration				
Surfactant	Intrapleural pressure left (atm)	Intrapleural pressure right (atm)	Airflow left (ml/min)	Airflow right (ml/min)	Total airflow (ml/min)

? PREDICT Question 1
What effect will adding more surfactant have on these lungs?

_____

6. Click **Surfactant** twice to dispense two more aliquots of the synthetic lipids and proteins onto the interior lining of the lungs.

7. Click **Start** to initiate breathing in the presence of additional surfactant and observe the tracing that develops.

8. Click **Record Data** to display your results in the grid (and record your results in Chart 3).

9. Click **Clear Tracings** to clear the tracing on the oscilloscope.

10. Click **Flush** to clear the lungs of surfactant from the previous run.

11. Click **Start** to initiate breathing and observe the tracing that develops. Notice the negative pressure condition displayed below the oscilloscope when the lungs inflate.

12. Click **Record Data** to display your results in the grid (and record your results in Chart 3).

13. Click the valve on the left side of the glass bell jar to open it.

14. Click **Start** to initiate breathing and observe the tracing that develops.

15. Click **Record Data** to display your results in the grid (and record your results in Chart 3).

? PREDICT Question 2
What will happen to the collapsed lung in the left side of the glass bell jar if you close the valve?

_____

16. Click the valve on the left side of the glass bell jar to close it.

17. Click **Start** to initiate breathing and observe the tracing that develops.

18. Click **Record Data** to display your results in the grid (and record your results in Chart 3).

19. Click the **Reset** button above the glass bell jar to draw the air out of the intrapleural space and return the lung to its normal resting condition.

20. Click **Start** to initiate breathing and observe the tracing that develops.

21. Click **Record Data** to display your results in the grid (and record your results in Chart 3).

After you complete the experiment, take the online **Post-lab Quiz** for Activity 3.

Activity Questions

1. Why is normal quiet breathing so difficult for premature infants?

_____

_____

2. Why does a pneumothorax frequently lead to atelectasis?

_____

_____

NAME_____

LAB TIME/DATE _____

Respiratory System Mechanics

ACTIVITY 1 Measuring Respiratory Volumes and Calculating Capacities

1. What would be an example of an everyday respiratory event the ERV button simulates? _____

2. What additional skeletal muscles are utilized in an ERV activity?

   _____

3. What was the FEV_1 (%) at the initial radius of 5.00 mm?

   _____

4. What happened to the FEV_1 (%) as the radius of the airways decreased? How well did the results compare with your prediction?

   _____

5. Explain why the results from the experiment suggest that there is an obstructive, rather than a restrictive, pulmonary problem.

   _____

ACTIVITY 2 Comparative Spirometry

1. What lung values changed (from those of the normal patient) in the spirogram when the patient with emphysema was selected? Why did these values change as they did? How well did the results compare with your prediction?

   _____

   _____

2. Which of these two parameters changed more for the patient with emphysema, the FVC or the FEV_1? _____

3. What lung values changed (from those of the normal patient) in the spirogram when the patient experiencing an acute asthma attack was selected? Why did these values change as they did? How well did the results compare with your prediction?

   _____

   _____

   _____

4. How is having an acute asthma attack similar to having emphysema? How is it different?_____

   _____

5. Describe the effect that the inhaler medication had on the asthmatic patient. Did all the spirogram values return to "normal"? Why do you think some values did not return all the way to normal? How well did the results compare with your prediction?

_____

_____

_____

6. How much of an increase in FEV_1 do you think is required for it to be considered significantly improved by the medication?

_____

7. With moderate aerobic exercise, which changed more from normal breathing, the ERV or the IRV? How well did the results compare with your prediction?

_____

_____

8. Compare the breathing rates during normal breathing, moderate exercise, and heavy exercise. _____

_____

ACTIVITY 3 Effect of Surfactant and Intrapleural Pressure on Respiration

1. What effect does the addition of surfactant have on the airflow? How well did the results compare with your prediction?

_____

_____

2. Why does surfactant affect airflow in this manner? _____

_____

3. What effect did opening the valve have on the left lung? Why does this happen?

_____

_____

4. What effect on the collapsed lung in the left side of the glass bell jar did you observe when you closed the valve? How well did the results compare with your prediction?

_____

_____

5. What emergency medical condition does opening the left valve simulate?

_____

6. In the last part of this activity, you clicked the Reset button to draw the air out of the intrapleural space and return the lung to its normal resting condition. What emergency procedure would be used to achieve this result if these were the lungs in a living person?

_____

_____

7. What do you think would happen when the valve is opened if the two lungs were in a single large cavity rather than separate

cavities? _____

_____

Chemical and Physical Processes of Digestion

PRE-LAB QUIZ

1. Circle the correct term: The <u>liver</u> / <u>stomach</u> produces pepsin which, in the presence of hydrochloric acid, digests protein.

2. _____ is a hydrolytic enzyme that breaks starch down to maltose.
   a. Pepsin
   b. Salivary amylase
   c. Pancreatic lipase
   d. Bile

3. A _____ is made in order to compare a known standard to an experimental standard.

4. Circle True or False: A positive IKI test for starch will yield a bright orange to green color.

5. An enzyme has a pocket or _____, which the substrate or substrates must fit into temporarily in order for catalysis to occur.
   a. Reagent
   b. Hydrolase
   c. Active site
   d. Polysaccharide

6. During digestion, the chief cells of the stomach secrete _____ which is responsible for the digestion of protein.
   a. Peptidase
   b. Amylase
   c. Lipase
   d. Pepsin

7. Circle the correct term: <u>Positive</u> / <u>Negative</u> controls are used to determine whether there are any contaminating substances in the reagents used in an experiment.

8. Circle True or False: At room temperature, both fats and oils are liquid and are soluble in water.

9. A solution containing fatty acids formed by lipase activity will have _____than a solution without such fatty acid production.
   a. A lower pH
   b. A higher pH
   c. A lower temperature
   d. A higher temperature

Exercise Overview

The **digestive system,** also called the gastrointestinal system, consists of the digestive tract (also called the gastrointestinal tract, or GI tract) and accessory glands that secrete enzymes and fluids needed for digestion. The digestive tract includes the mouth, pharynx, esophagus, stomach, small intestine, colon, rectum, and anus. The major functions of the digestive system are to ingest food, to break food down to its simplest components, to extract nutrients from these components for absorption into the body, and to eliminate wastes.

Most of the food we consume cannot be absorbed into our bloodstream without first being broken down into smaller subunits. **Digestion** is the process of

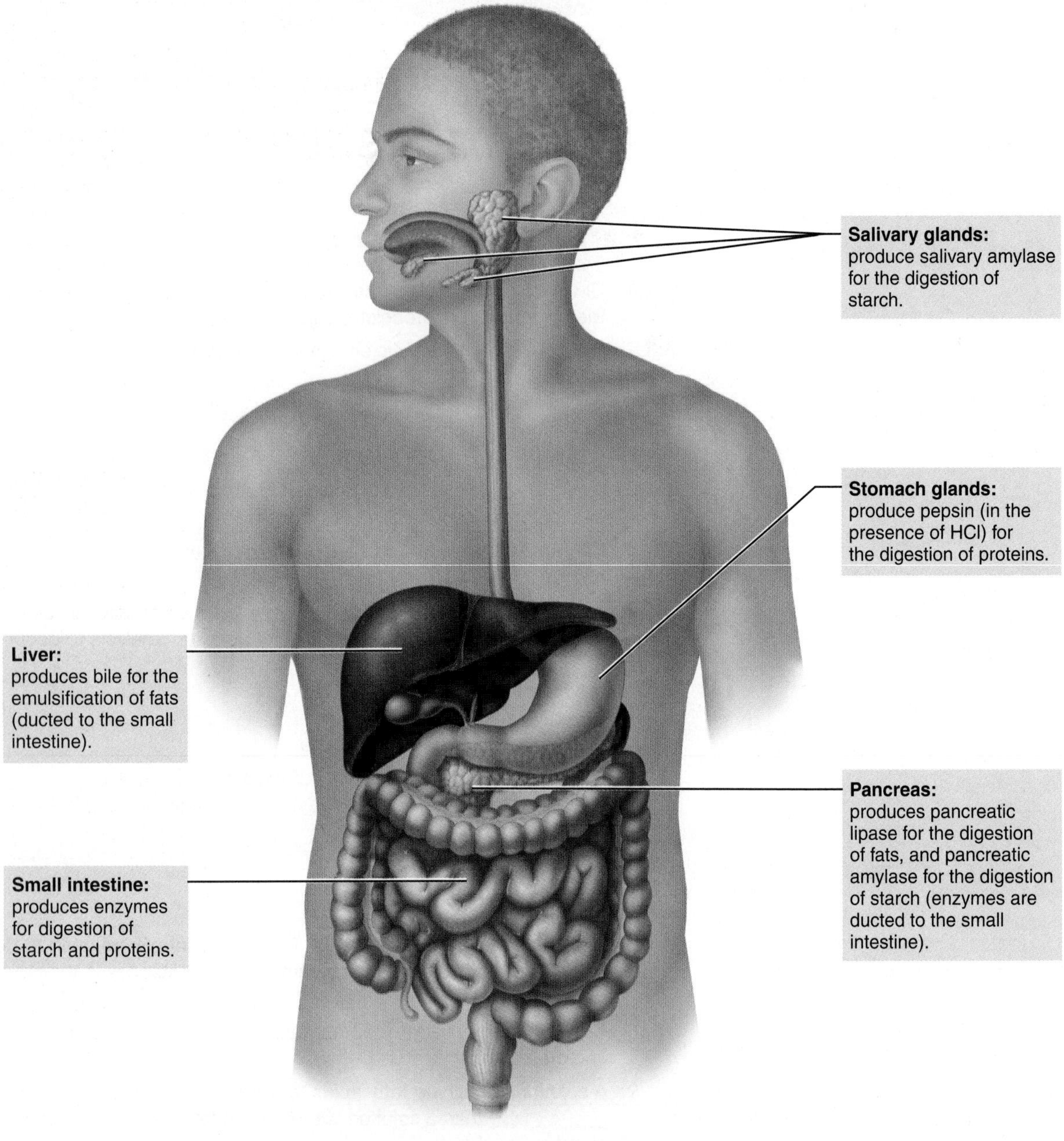

FIGURE 8.1   The human digestive system. A few sites of chemical digestion and the organs that produce the enzymes of chemical digestion.

breaking down food molecules into smaller molecules with the aid of enzymes in the digestive tract. **Enzymes** are large protein molecules produced by body cells. They are biological catalysts that increase the rate of a chemical reaction without becoming part of the product. The digestive enzymes are hydrolytic enzymes, or **hydrolases,** which break down organic food molecules, or **substrates,** by adding water to the molecular bonds, thus cleaving the bonds between the subunits, or monomers.

A hydrolytic enzyme is highly specific in its action. Each enzyme hydrolyzes one substrate molecule or, at most, a small group of substrate molecules. Specific environmental conditions are necessary for an enzyme to function optimally. For example, in extreme environments, such as high temperature, an enzyme can unravel, or denature, because of the effect that temperature has on the three-dimensional structure of the protein.

Because digestive enzymes actually function outside the body cells in the digestive tract lumen, their hydrolytic activity can also be studied in vitro in a test tube. Such in vitro studies provide a convenient laboratory environment for investigating the effect of various factors on enzymatic activity. View Figure 8.1 for an overview of chemical digestion sites in the body.

Assessing Starch Digestion by Salivary Amylase

OBJECTIVES

1. To explain how enzyme activity can be assessed with enzyme assays: the IKI assay and the Benedict's assay.

2. To define enzyme, catalyst, hydrolase, substrate, and control.

3. To understand the specificity of amylase action.

4. To name the end products of carbohydrate digestion.

5. To perform the appropriate chemical tests to determine whether digestion of a particular food has occurred.

6. To discuss the possible effect of temperature and pH on amylase activity.

Introduction

In this activity you will investigate the hydrolysis of starch to maltose by **salivary amylase,** the enzyme produced by the salivary glands and secreted into the mouth. For you to be able to detect whether or not enzymatic action has occurred, you need to be able to identify the presence of the substrate and the product to determine to what extent hydrolysis has occurred. Thus, **controls** must be prepared to provide a known standard against which comparisons can be made. With positive controls, all of the required substances are included and a positive result is expected. Sometimes negative controls are included. With negative controls, a negative result is expected. Negative results with negative controls validate the experiment. Negative controls are used to determine whether there are any contaminating substances in the reagents. So, when a positive result is produced but a negative result is expected, one or more contaminating substances are present to cause the change.

With amylase activity, starch decreases and maltose increases as digestion proceeds according to the following equation.

$$\text{Starch} + \text{water} \xrightarrow{\text{amylase}} \text{maltose}$$

Because the chemical changes that occur as starch is digested to maltose cannot be seen by the naked eye, you need to conduct an **enzyme assay,** the chemical method of detecting the presence of digested substances. You will perform two enzyme assays on each sample. The IKI assay detects the presence of starch, and the Benedict's assay tests for the presence of reducing sugars, such as glucose or maltose, which are the digestion products of starch. Normally a caramel-colored solution, IKI turns blue-black in the presence of starch. Benedict's reagent is a bright blue solution that changes to green to orange to reddish brown with increasing amounts of maltose. It is important to understand that enzyme assays only indicate the presence or absence of substances. It is up to you to analyze the results of the experiments to decide whether enzymatic hydrolysis has occurred.

EQUIPMENT USED The following equipment will be depicted on-screen: amylase—an enzyme that digests starch; starch—a complex carbohydrate substrate; maltose—a disaccharide substrate; pH buffers—solutions used to adjust the pH of the solution; deionized water—used to adjust the volume so that it is the same for each reaction; test tubes—used as reaction vessels for the various tests; incubators—used for temperature treatments (boiling, freezing, and 37°C incubation); IKI—found in the assay cabinet; used to detect the presence of starch; Benedict's reagent—found in the assay cabinet; used to detect the products of starch digestion (this includes the reducing sugars maltose and glucose).

Experiment Instructions

Go to the home page in the PhysioEx software and click **Exercise 8: Chemical and Physical Processes of Digestion.** Click **Activity 1: Assessing Starch Digestion by Salivary Amylase,** and take the online **Pre-lab Quiz** for Activity 1.

After you take the online Pre-lab Quiz, click the **Experiment** tab and begin the experiment. The experiment instructions are reprinted here for your reference. The opening screen for the experiment is shown below.

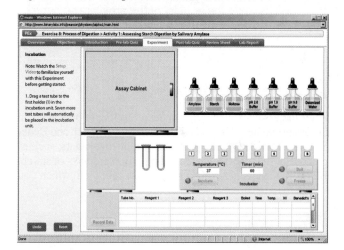

Incubation

1. Drag a test tube to the first holder (**1**) in the incubation unit. Seven more test tubes will automatically be placed in the incubation unit.

2. Add the substances indicated below to tubes 1 through 7.

Tube 1: amylase, starch, pH 7.0 buffer

Tube 2: amylase, starch, pH 7.0 buffer

Tube 3: amylase, starch, pH 7.0 buffer

Tube 4: amylase, deionized water, pH 7.0 buffer

Tube 5: deionized water, starch, pH 7.0 buffer

Tube 6: deionized water, maltose, pH 7.0 buffer

Tube 7: amylase, starch, pH 2.0 buffer

Tube 8: amylase, starch, pH 9.0 buffer

To add a substance to a test tube, drag the dropper cap of the bottle on the solutions shelf to the top of the test tube.

3. Click the number (**1**) under the first test tube. The tube will descend into the incubation unit. All other tubes should remain in the raised position.

4. Click **Boil** to boil tube 1. After boiling for a few moments, the tube will automatically rise.

5. Click the number (**2**) under the second test tube. The tube will descend into the incubation unit. All other tubes should remain in the raised position.

6. Click **Freeze** to freeze tube 2. After freezing for a few moments, the tube will automatically rise.

7. Click **Incubate** to start the run. Note that the incubation temperature is set at 37°C and the timer is set at 60 min. The incubation unit will gently agitate the test tube rack, evenly mixing the contents of all test tubes throughout the incubation. The simulation compresses the 60-minute time period into 10 seconds of real time, so what would be a 60-minute incubation in real time will take only 10 seconds in the simulation. When the incubation time elapses, the test tube rack will automatically rise, and the doors to the assay cabinet will open.

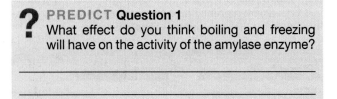

PREDICT Question 1
What effect do you think boiling and freezing will have on the activity of the amylase enzyme?

_____

_____

Assays

After the assay cabinet doors open, notice the two reagents in the assay cabinet. IKI tests for the presence of starch and Benedict's reagent detects the presence of reducing sugars, such as glucose or maltose, which are the digestion products of starch. Below the reagents are eight small assay tubes into which you

will dispense a small amount of test solution from the incubated samples in the incubation unit, plus a drop of IKI.

8. Drag the first tube in the incubation unit to the first small assay tube on the left side of the assay cabinet to decant approximately half of the contents in the test tube into the assay tube. The decanting step will automatically repeat for the remaining tubes in the incubation unit.

9. Drag the IKI dropper cap to the first assay tube to dispense a drop of IKI into the assay tube. The dropper will automatically move across and dispense IKI to the remaining tubes.

10. Inspect the tubes for color change. A blue-black color indicates a positive starch test. If starch is not present, the mixture will look like diluted IKI, a negative starch test. Intermediate starch amounts result in a pale-gray color. Click **Record Data** to display your results in the grid (and record your results in Chart 1).

11. Drag the Benedict's reagent dropper cap to the test tube in the first holder (**1**) in the incubation unit to dispense five drops of Benedict's reagent into the tube. The dropper will automatically move across and dispense Benedict's reagent to the remaining tubes.

12. Click **Boil.** The entire tube rack will descend into the incubation unit and automatically boil the tube contents for a few moments.

13. Inspect the tubes for color change. A green-to-reddish color indicates that a reducing sugar is present; this is a positive sugar test. An orange-colored sample contains more sugar than a green sample. A reddish-brown color indicates even more sugar. A negative sugar test is indicated by no color change from the original bright blue. Click **Record Data** to display your results in the grid (and record your results in Chart 1).

After you complete the experiment, take the online **Post-lab Quiz** for Activity 1.

CHART 1	Salivary Amylase Digestion of Starch							
Tube No.	1	2	3	4	5	6	7	8
Additives	Amylase Starch pH 7.0 buffer	Amylase Starch pH 7.0 buffer	Amylase Starch pH 7.0 buffer	Amylase Deionized water pH 7.0 buffer	Deionized water Starch pH 7.0 buffer	Deionized water Maltose pH 7.0 buffer	Amylase Starch pH 2.0 buffer	Amylase Starch pH 9.0 buffer
Incubation condition	Boil first, then incubate at 37°C for 60 minutes	Freeze first, then incubate at 37°C for 60 minutes	37°C 60 minutes	37°C 60 minutes	37°C 60 minutes	37°C 60 minutes	37°C 60 minutes	37°C 60 minutes
IKI test								
Benedict's test								

Activity Questions

1. Describe the effect that boiling had on the activity of amylase. Why did boiling have this effect? How does the effect of freezing differ from the effect of boiling?

_____

_____

2. What is the purpose for including tube 3 and what can you conclude from the result?

_____

_____

3. Describe how you determined the optimal pH for amylase activity.

_____

_____

4. Judging from what you learned in this activity, suggest a reason why salivary amylase would be much less active in the stomach.

_____

_____

Exploring Amylase Substrate Specificity

OBJECTIVES

1. Explain how hydrolytic enzyme activity can be assessed with the IKI assay and the Benedict's assay.
2. Understand the specificity that enzymes have for their substrate.
3. Understand the difference between the substrates starch and cellulose.
4. Explain what would be the substrate specificity of peptidase.
5. Explain how bacteria might aid in digestion.

Introduction

In this activity you will investigate the specificity that enzymes have for their substrates. To do this you will hydrolyze starch to maltose and maltotriose using **salivary amylase,** the enzyme produced by the salivary glands and secreted into the mouth. To detect whether or not enzymatic action has occurred, you need to be able to identify the presence of the substrate and the product to determine to what extent hydrolysis has occurred. The **substrate** is the substance that the enzyme acts on. The enzyme has a pocket called the **active site,** which the substrate or substrates must fit into temporarily for catalysis to occur. The substrate is often held in the active site by non-covalent bonds (weak bonds), such as ionic bonds and hydrogen bonds.

With amylase activity, starch decreases and sugar increases as digestion proceeds according to the following equation.

$$\text{Starch} + \text{water} \xrightarrow{\text{amylase}} \text{maltose} + \text{maltotriose} + \text{starch}$$

Because the chemical changes that occur as starch is digested to maltose cannot be seen by the naked eye, you need to conduct an **enzyme assay,** the chemical method of detecting the presence of digested substances. You will perform two enzyme assays on each sample. The IKI assay detects the presence of starch or cellulose and the Benedict's assay tests for the presence of reducing sugars, such as glucose or maltose, which are the digestion products of starch. Normally a caramel-colored solution, IKI turns blue-black in the presence of starch or cellulose. Benedict's reagent is a bright blue solution that changes to green to orange to reddish brown with increasing amounts of maltose. It is important to understand that enzyme assays only indicate the presence or absence of substances. It is up to you to analyze the results of the experiments to decide whether enzymatic hydrolysis has occurred.

Starch is a polysaccharide found in plants, where it is used to store energy. Plants also have the polysaccharide **cellulose,** which provides rigidity to their cell walls. Both polysaccharides are polymers of glucose, but the glucose molecules are linked differently. You will be testing salivary amylase to determine whether it digests cellulose. Also, you will investigate to see whether a bacterial suspension can digest cellulose and whether **peptidase,** a pancreatic enzyme that digests peptides, can break down starch.

> **EQUIPMENT USED** The following equipment will be depicted on-screen: amylase—an enzyme that digests starch; starch—a polysaccharide; pH 7.0 buffer—a solution used to set the pH of the test tube solution; deionized water—used to adjust the test tube solution volume so it is the same for each reaction; glucose—a reducing sugar that is the monosaccharide subunit of both starch and cellulose; cellulose—a complex carbohydrate found in the cell wall of plants; peptidase—a pancreatic enzyme that breaks down peptides; bacteria—a suspension of live bacteria; test tubes—used as reaction vessels for the various tests; incubators—used for temperature treatments (37°C incubation); IKI—found in the assay cabinet; used to detect the presence of starch or cellulose; Benedict's reagent—found in the assay cabinet; used to detect the products of starch and cellulose digestion.

Experiment Instructions

Go to the home page in the PhysioEx software and click **Exercise 8: Chemical and Physical Processes of Digestion.** Click **Activity 2: Exploring Amylase Substrate Specificity,** and take the online **Pre-lab Quiz** for Activity 2.

After you take the online Pre-lab Quiz, click the **Experiment** tab and begin the experiment. The experiment instructions are reprinted here for your reference. The opening screen for the experiment is shown on the following page.

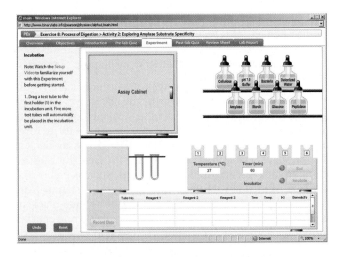

Incubation

1. Drag a test tube to the first holder (**1**) in the incubation unit. Five more test tubes will automatically be placed in the incubation unit.

2. Add the substances indicated below to tubes 1 through 6.

Tube 1: amylase, starch, pH 7.0 buffer

Tube 2: amylase, glucose, pH 7.0 buffer

Tube 3: amylase, cellulose, pH 7.0 buffer

Tube 4: cellulose, pH 7.0 buffer, deionized water

Tube 5: peptidase, starch, pH 7.0 buffer

Tube 6: bacteria, cellulose, pH 7.0 buffer

To add a substance to a test tube, drag the dropper cap of the bottle on the solutions shelf to the test tube.

3. Click **Incubate** to start the run. Note that the incubation temperature is set at 37°C and the timer is set at 60 min. The incubation unit will gently agitate the test tube rack, evenly mixing the contents of all test tubes throughout the incubation. The simulation compresses the 60-minute time period into 10 seconds of real time, so what would be a 60-minute incubation in real life will take only 10 seconds in the simulation. When the incubation time elapses, the test tube rack will automatically rise, and the doors to the assay cabinet will open.

? PREDICT Question 1
Do you think test Tube 3 will show a positive Benedict's test?

_____

_____

Assays

After the assay cabinet doors open, notice the two reagents in the assay cabinet. IKI tests for the presence of starch and Benedict's reagent detects the presence of reducing sugars, such as glucose or maltose, which are the digestion products of starch. Below the reagents are seven small assay tubes into which you will dispense a portion of the incubated samples, plus a drop of IKI.

4. Drag the first tube in the incubation unit to the first small assay tube on the left side of the assay cabinet to decant approximately half of the contents in the test tube into the assay tube. The decanting step will automatically repeat for the remaining tubes in the incubation unit.

5. Drag the IKI dropper cap to the first assay tube to dispense a drop of IKI into the assay tube. The dropper will automatically dispense IKI into the remaining tubes.

6. Inspect the tubes for color change. A blue-black color indicates a positive starch test. If starch is not present, the mixture will look like diluted IKI, a negative starch test. Intermediate starch amounts result in a pale-gray color. Click **Record Data** to display your results in the grid (and record your results in Chart 2).

7. Drag the Benedict's reagent dropper cap to the test tube in the first holder (**1**) in the incubation unit to dispense five drops of Benedict's reagent into the tube. The dropper will automatically move across and dispense Benedict's reagent to the remaining tubes.

8. Click **Boil**. The entire tube rack will descend into the incubation unit and automatically boil the tube contents for a few moments.

9. Inspect the tubes for color change. A green-to-reddish color indicates that a reducing sugar is present; this is a positive sugar test. An orange-colored sample contains more sugar than a green sample. A reddish-brown color indicates even more sugar. A negative sugar test is indicated by no color change from the original bright blue. Click **Record Data** to display your results in the grid (and record your results in Chart 2).

CHART 2	Enzyme Digestion of Starch and Cellulose					
Tube No.	1	2	3	4	5	6
Additives	Amylase Starch pH 7.0 buffer	Amylase Glucose pH 7.0 buffer	Amylase Cellulose pH 7.0 buffer	Deionized water Cellulose pH 7.0 buffer	Peptidase Starch pH 7.0 buffer	Bacteria Cellulose pH 7.0 buffer
Incubation condition	37°C 60 minutes	37°C 60 minutes	37°C 60 minutes	37°C 60 minutes	37°C 60 minutes	37°C 60 minutes
IKI test						
Benedict's test						

After you complete the Experiment, take the online **Post-lab Quiz** for Activity 2.

Activity Questions

1. Does amylase use cellulose as a substrate?

_____

_____

2. What effect did the addition of bacteria have on the digestion of cellulose?

_____

_____

3. What effect did the addition of peptidase to the starch have? Why?

_____

_____

4. What is the smallest subunit into which starch can be broken down?

_____

_____

_____ ▬

Assessing Pepsin Digestion of Protein

OBJECTIVES

1. Explain how the enzyme activity of pepsin can be assessed with the BAPNA assay.
2. Identify the substrate specificity of pepsin.
3. Discuss the effects of temperature and pH on pepsin activity.
4. Understand the pH specificity of enzyme activity and how it relates to human physiology.

Introduction

In this activity, you will explore the digestion of protein **(peptides).** Peptides are two or more **amino acids** linked together by a peptide bond. A peptide chain containing 10 to 100 amino acids is typically called a **polypeptide. Proteins** can consist of a large peptide chain (more than 100 amino acids) or even multiple peptide chains.

During digestion, **chief cells** of the stomach glands secrete a protein-digesting enzyme called **pepsin.** Pepsin **hydrolyzes** peptide bonds. This activity breaks up ingested proteins and polypeptides into smaller peptide chains and free amino acids. In this activity, you will use **BAPNA** as a **substrate** to assess pepsin activity. BAPNA is a synthetic "peptide" that releases a yellow dye **product** when hydrolyzed. BAPNA solutions turn yellow in the presence of an active peptidase, such as pepsin, but otherwise remain colorless.

To quantify the pepsin activity in each test solution, you will use a **spectrophotometer** to measure the amount of yellow dye produced. A spectrophotometer shines light through the sample and then measures how much light is absorbed. The fraction of light absorbed is expressed as the sample's **optical density.** Yellow solutions, where BAPNA has been hydrolyzed, will have optical densities greater than zero. The greater the optical density, the more hydrolysis has occurred. Colorless solutions, in contrast, do not absorb light and will have an optical density near zero.

Some negative controls are included in this activity. With negative controls, a negative result is expected. Negative results with negative controls validate the experiment. Negative controls are used to determine whether there are any contaminating substances in the reagents. So, when a positive result is produced but a negative result is expected, one or more contaminating substances are present to cause the change.

> **EQUIPMENT USED** The following equipment will be depicted on-screen: pepsin—an enzyme that digests peptides; BAPNA—a synthetic "peptide"; pH buffers—solutions used to set the pH of the test tube solution; deionized water—used to adjust the test tube solution volume so it is the same for each reaction; test tubes—used as reaction vessels for the various tests; incubators—used for temperature treatments (boiling and 37°C incubation); spectrophotometer—found in the assay cabinet; used to measure the optical density of solutions.

Experiment Instructions

Go to the home page in the PhysioEx software and click **Exercise 8: Chemical and Physical Processes of Digestion.** Click **Activity 3: Assessing Pepsin Digestion of Protein,** and take the online **Pre-lab Quiz** for Activity 3.

After you take the online Pre-lab Quiz, click the **Experiment** tab and begin the experiment. The experiment instructions are reprinted here for your reference. The opening screen for the experiment is shown below.

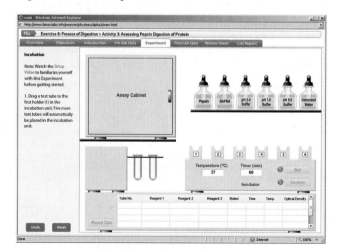

Incubation

1. Drag a test tube to the first holder (**1**) in the incubation unit. Five more test tubes will automatically be placed in the incubation unit.

2. Add the substances indicated below to tubes 1 through 6.

Tube 1: pepsin, BAPNA, pH 2.0 buffer

Tube 2: pepsin, BAPNA, pH 2.0 buffer

Tube 3: pepsin, deionized water, pH 2.0 buffer

Tube 4: deionized water, BAPNA, pH 2.0 buffer

Tube 5: pepsin, BAPNA, pH 7.0 buffer

Tube 6: pepsin, BAPNA, pH 9.0 buffer

To add a substance to a test tube, drag the dropper cap of the bottle on the solutions shelf to the test tube.

3. Click the number (1) under the first test tube. The tube will descend into the incubation unit. All other tubes should remain in the raised position.

4. Click **Boil** to boil tube 1. After boiling for a few moments, the tube will automatically rise.

5. Click **Incubate** to start the run. Note that the incubation temperature is set at 37°C and the timer is set at 60 min. The incubation unit will gently agitate the test tube rack, evenly mixing the contents of all test tubes throughout the incubation. The simulation compresses the 60-minute time period into 10 seconds of real time, so what would be a 60-minute incubation in real life will take only 10 seconds in the simulation. When the incubation time elapses, the test tube rack will automatically rise, and the doors to the assay cabinet will open. The spectrophotometer is in the assay cabinet.

> **? PREDICT Question 1**
> At which pH do you think pepsin will have the highest activity?
>
> _____
>
> _____

Assays

6. You will now use the spectrophotometer to measure how much yellow dye was liberated from BAPNA hydrolysis. Drag the first tube in the incubation unit to the holder in the spectrophotometer to drop the tube into the holder.

7. Click **Analyze.** The spectrophotometer will shine light through the solution to measure the amount of light absorbed, which it reports as the solution's optical density. The optical density of the sample is shown in the optical density display.

8. Click **Record Data** to display your results in the grid (and record your results in Chart 3).

9. Drag the tube to its original position in the incubation unit.

10. Analyze the remaining five tubes by repeating the following steps for each tube.

- Drag the tube to the holder in the spectrophotometer to drop the tube into the holder.
- Click **Analyze.**
- Drag the tube to its original position in the incubation unit.

After you have analyzed all five tubes, click **Record Data** to display your results in the grid. (and record your results in Chart 3).

After you complete the experiment, take the online **Post-lab Quiz** for Activity 3.

Activity Questions

1. Describe the significance of the optimum pH for pepsin observed in the simulation and the secretion of pepsin by the chief cells of the gastric glands.

_____

_____

2. Would pepsin be active in the mouth? Explain your answer.

_____

_____

3. What are the subunit products of peptide digestion?

_____

_____

4. Describe the reason for including control tube 4.

_____

_____

CHART 3	Pepsin Digestion of Protein					
Tube No.	1	2	3	4	5	6
Additives	Pepsin BAPNA pH 2.0 buffer	Pepsin BAPNA pH 2.0 buffer	Pepsin Deionized water pH 2.0 buffer	Deionized water BAPNA pH 2.0 buffer	Pepsin BAPNA pH 7.0 buffer	Pepsin BAPNA pH 9.0 buffer
Incubation condition	Boil first, then incubate at 37°C for 60 minutes	37°C 60 minutes	37°C 60 minutes	37°C 60 minutes	37°C 60 minutes	37°C 60 minutes
Optical density						

Assessing Lipase Digestion of Fat

OBJECTIVES

1. Explain how the enzyme activity of pancreatic lipase can be assessed with a pH-based measurement.
2. Identify the hydrolysis products of fat digestion.
3. Understand the role that bile plays in fat digestion.
4. Understand the significance of pH specificity of lipase activity and how it relates to human physiology.
5. Discuss the difficulty of using pH to measure digestion when comparing the activity of lipase at various pHs.

Introduction

Fats and oils belong to a diverse class of molecules called lipids. **Triglycerides,** a type of lipid, make up both fats and oils. At room temperature, fats are solid and oils are liquid. Both are poorly soluble in water. This insolubility of triglycerides presents a challenge during digestion because they tend to clump together, leaving only the surface molecules exposed to **lipase** enzymes. To overcome this difficulty, **bile salts** are secreted into the small intestine during digestion to physically emulsify lipids. Bile salts act like a detergent, separating the lipid clumps and increasing the surface area accessible to lipase enzymes.

As a result, two reactions must occur. First,

$$\text{Triglyceride clumps} \xrightarrow[\text{(emulsification)}]{\text{bile}} \text{minute triglyceride droplets}$$

Then,

$$\text{Triglyceride} \xrightarrow{\text{lipase}} \text{monoglyceride} + \text{two fatty acids}$$

Lipase hydrolyzes each triglyceride to a monoglyceride and two fatty acids. In addition to the **pancreatic lipase** secreted into the small intestine, **lingual lipase** and **gastric lipase** are also secreted. Even though bile salts are not secreted in the mouth or the stomach, small amounts of lipids are digested by these other lipases.

Because some of the end products of fat digestion are acidic (that is, fatty acids), lipase activity can be easily measured by monitoring the solution's **pH.** A solution containing fatty acids liberated by lipase activity will have a lower pH than a solution without such fatty acid production. You will record pH in this activity with a **pH meter.**

EQUIPMENT USED The following equipment will be depicted on-screen: lipase—an enzyme that digests triglycerides; vegetable oil—a mixture of triglycerides; bile salts—a solution that physically separates fats into smaller droplets; pH buffers—solutions used to set the pH of the test tube solution; deionized water—used to adjust the test tube solution volume so it is the same for each reaction; test tubes—used as reaction vessels for the various tests; incubators—used for temperature treatments (boiling and 37°C incubation); pH meter—found in the assay cabinet; used to measure pH.

Experiment Instructions

Go to the home page in the PhysioEx software and click **Exercise 8: Chemical and Physical Processes of Digestion.** Click **Activity 4: Assessing Lipase Digestion of Fat,** and take the online **Pre-lab Quiz** for Activity 4.

After you take the online Pre-lab Quiz, click the **Experiment** tab and begin the experiment. The experiment instructions are reprinted here for your reference. The opening screen for the experiment is shown below.

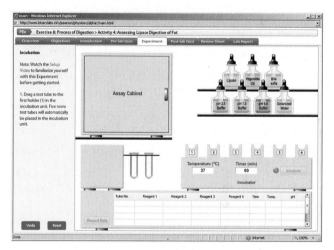

Incubation

1. Drag a test tube to the first holder (**1**) in the incubation unit. Five more test tubes will automatically be placed in the incubation unit.

2. Add the substances indicated below to tubes 1 through 6.

Tube 1: lipase, vegetable oil, bile salts, pH 7.0 buffer

Tube 2: lipase, vegetable oil, deionized water, pH 7.0 buffer

Tube 3: lipase, deionized water, bile salts, pH 9.0 buffer

Tube 4: deionized water, vegetable oil, bile salts, pH 7.0 buffer

Tube 5: lipase, vegetable oil, bile salts, pH 2.0 buffer

Tube 6: lipase, vegetable oil, bile salts, pH 9.0 buffer

To add a substance to a test tube, drag the dropper cap of the bottle on the solutions shelf to the test tube.

3. Click **Incubate** to start the run. Note that the incubation temperature is set at 37°C and the timer is set at 60 min. The incubation unit will gently agitate the test tube rack, evenly mixing the contents of all test tubes throughout the incubation. The simulation compresses the 60-minute time period into 10 seconds of real time, so what would be a 60-minute incubation in real life will take only 10 seconds in the simulation. When the incubation time elapses, the test tube rack will automatically rise, and the doors to the assay cabinet will open.

PREDICT Question 1
Which tube do you think will have the highest lipase activity?

_____

_____

Assays

4. After the assay cabinet doors open, you will see a pH meter that you will use to measure the final pH of your test solutions. Drag the first tube in the incubation unit to the holder in the pH meter to drop the tube into the holder.

5. Click **Measure pH.** A probe will descend into the sample, take a pH reading, and then retract.

6. Click **Record Data** to display your results in the grid (and record your results in Chart 4).

7. Drag the tube to its original position in the incubation unit.

8. Measure the pH in the remaining five tubes by repeating the following steps for each tube.

- Drag the tube in the incubation unit to the holder in the pH meter to drop the tube into the holder.
- Click **Measure pH.**
- Drag the tube to its original position in the incubation unit.

After you have measured the pH in all five tubes, click **Record Data** to display your results in the grid (and record your results in Chart 4).

After you complete the experiment, take the online **Post-lab Quiz** for Activity 4.

Activity Questions

1. Describe how lipase activity is measured in the simulation.

_____

_____

2. Can you determine if fat hydrolysis occurred in tube 5? Why or why not?

_____

_____

3. Would pancreatic lipase be active in the mouth? Why or why not?

_____

_____

4. Describe the physical separation of fats by bile salts.

_____

_____

CHART 4	Pancreatic Lipase Digestion of Triglycerides and the Action of Bile					
Tube No.	1	2	3	4	5	6
Additives	Lipase Vegetable oil Bile salts pH 7.0 buffer	Lipase Vegetable oil Deionized water pH 7.0 buffer	Lipase Deionized water Bile salts pH 9.0 buffer	Deionized water Vegetable oil Bile salts pH 7.0 buffer	Lipase Vegetable oil Bile salts pH 2.0 buffer	Lipase Vegetable oil Bile salts pH 9.0 buffer
Incubation condition	37°C 60 minutes	37°C 60 minutes	37°C 60 minutes	37°C 60 minutes	37°C 60 minutes	37°C 60 minutes
pH						

NAME_____

LAB TIME/DATE_____

Chemical and Physical Processes of Digestion

ACTIVITY 1   Assessing Starch Digestion by Salivary Amylase

1. List the substrate and the subunit product of amylase. _____

   _____

2. What effect did boiling and freezing have on enzyme activity? Why? How well did the results compare with your prediction?

   _____

   _____

3. At what pH was the amylase most active? Describe the significance of this result. _____

   _____

4. Briefly describe the need for controls and give an example used in this activity. _____

   _____

5. Describe the significance of using a 37°C incubation temperature to test salivary amylase activity. _____

   _____

ACTIVITY 2   Exploring Amylase Substrate Specificity

1. Describe why the results in tube 1 and tube 2 are the same. _____

   _____

2. Describe the result in tube 3. How well did the results compare with your prediction? _____

   _____

3. Describe the usual substrate for peptidase. _____

   _____

4. Explain how bacteria can aid in digestion. _____

   _____

ACTIVITY 3 Assessing Pepsin Digestion of Protein

1. Describe the effect that boiling had on pepsin and how you could tell that it had that effect. _____

_____

2. Was your prediction correct about the optimal pH for pepsin activity? Discuss the physiological correlation behind your results.

_____

_____

3. What do you think would happen if you reduced the incubation time to 30 minutes for tube 5? _____

_____

ACTIVITY 4 Assessing Lipase Digestion of Fat

1. Explain why you can't fully test the lipase activity in tube 5. _____

_____

2. Which tube had the highest lipase activity? How well did the results compare with your prediction? Discuss possible reasons

why it may or may not have matched. _____

_____

_____

3. Explain why pancreatic lipase would be active in both the mouth and the intestine. _____

_____

_____

4. Describe the process of bile emulsification of lipids and how it improves lipase activity. _____

_____

_____

Renal System Physiology

Exercise Overview

The **kidney** is *both* an excretory and a regulatory organ. By filtering the water and solutes in the blood, the kidneys are able to *excrete* excess water, waste products, and even foreign materials from the body. However, the kidneys also *regulate* (1) plasma osmolarity (the concentration of a solution expressed as osmoles of solute per liter of solvent), (2) plasma volume, (3) the body's acid-base balance, and (4) the body's electrolyte balance. All these activities are extremely important for maintaining homeostasis in the body.

The paired kidneys are located between the posterior abdominal wall and the abdominal peritoneum. The right kidney is slightly lower than the left kidney. Each human kidney contains approximately one million **nephrons,** the functional units of the kidney.

Each nephron is composed of a **renal corpuscle** and a **renal tubule.** The renal corpuscle consists of a "ball" of capillaries, called the *glomerulus,* which is enclosed by a fluid-filled capsule, called *Bowman's capsule,* or the glomerular capsule. An **afferent arteriole** supplies blood to the glomerulus. As blood flows through the glomerular capillaries, protein-free plasma filters into the Bowman's capsule, a process called **glomerular filtration.** An **efferent arteriole** then drains the glomerulus of the remaining blood (view Figure 9.1).

The filtrate flows from Bowman's capsule into the start of the renal tubule, called the **proximal convoluted tubule,** then into the **loop of Henle,** a U-shaped hairpin loop, and, finally, into the **distal convoluted tubule** before emptying into

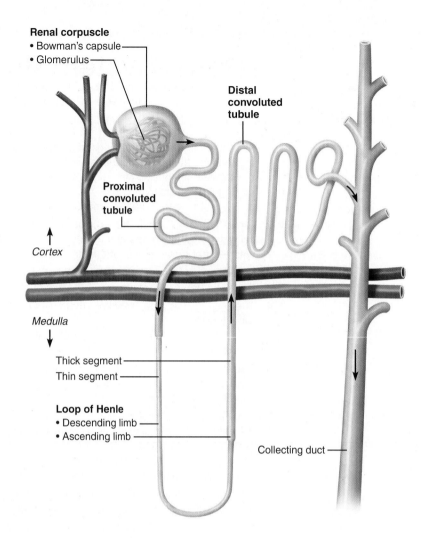

Renal corpuscle
• Bowman's capsule
• Glomerulus

Distal convoluted tubule

Proximal convoluted tubule

Cortex

Medulla

Thick segment
Thin segment

Loop of Henle
• Descending limb
• Ascending limb

Collecting duct

FIGURE 9.1   Location and structure of nephrons.

a **collecting duct**. From the collecting duct, the filtrate flows into, and collects in, the minor calyces.

The nephron performs three important functions that process blood into filtrate and urine: (1) glomerular filtration, (2) tubular reabsorption, and (3) tubular secretion. **Glomerular filtration** is a passive process in which fluid passes from the lumen of the glomerular capillary into the glomerular capsule of the renal tubule. **Tubular reabsorption** moves most of the filtrate back into the blood, leaving mainly salt water and the wastes in the lumen of the tubule. Some of the desirable, or needed, solutes are actively reabsorbed, and others move passively from the lumen of the tubule into the interstitial spaces. **Tubular secretion** is essentially the reverse of tubular reabsorption and is a process by which the kidneys can rid the blood of additional unwanted substances, such as creatinine and ammonia.

The reabsorbed solutes and water that move into the interstitial space between the nephrons need to be returned to the blood, or the kidneys will rapidly swell like balloons. The **peritubular capillaries** surrounding the renal tubule reclaim the reabsorbed substances and return them to general circulation. Peritubular capillaries arise from the efferent arteriole exiting the glomerulus and empty into the renal veins leaving the kidney.

ACTIVITY 1

The Effect of Arteriole Radius on Glomerular Filtration

OBJECTIVES

1. To understand the terms *nephron, glomerulus, glomerular capillaries, renal tubule, filtrate, Bowman's capsule, renal corpuscle, afferent arteriole, efferent arteriole, glomerular capillary pressure,* and *glomerular filtration rate.*

2. To understand how changes in afferent arteriole radius impact glomerular capillary pressure and filtration.

3. To understand how changes in efferent arteriole radius impact glomerular capillary pressure and filtration.

Introduction

Each of the million **nephrons** in each kidney contains two major parts: (1) a tubular component, the **renal tubule,** and (2) a vascular component, the **renal corpuscle** (view Figure 9.1). The **glomerulus** is a tangled capillary knot that filters fluid from the blood into the lumen of the renal tubule. The function of the renal tubule is to process the filtered fluid, also called the **filtrate.** The beginning of the renal tubule is an enlarged end called **Bowman's capsule** (or the glomerular capsule), which surrounds the glomerulus and serves to funnel the filtrate into the rest of the renal tubule. Collectively, the glomerulus and Bowman's capsule are called the renal corpuscle.

Two arterioles are associated with each glomerulus: an **afferent arteriole** feeds the **glomerular capillary** bed and an **efferent arteriole** drains it. These arterioles are responsible for blood flow through the glomerulus. The diameter of the efferent arteriole is smaller than the diameter of the afferent arteriole, restricting blood flow out of the glomerulus. Consequently, the pressure in the glomerular capillaries forces fluid through the endothelium of the capillaries into the lumen of the surrounding Bowman's capsule. In essence, everything in the blood except for the blood cells (red and white) and plasma proteins is filtered through the glomerular wall. From the Bowman's capsule, the filtrate moves into the rest of the renal tubule for processing. The job of the tubule is to reabsorb all the beneficial substances from its lumen and allow the wastes to travel down the tubule for elimination from the body.

During glomerular filtration, blood enters the glomerulus from the afferent arteriole and protein-free plasma flows from the blood across the walls of the glomerular capillaries and into the Bowman's capsule. The **glomerular filtration rate** is an index of kidney function. In humans, the filtration rate ranges from 80 to 140 ml/min, so that, in 24 hours, as much as 180 liters of filtrate is produced by the glomeruli. The filtrate formed is devoid of cellular debris, is essentially protein free, and contains a concentration of salts and organic molecules similar to that in blood.

The glomerular filtration rate can be altered by changing arteriole resistance or arteriole hydrostatic pressure. In this activity, you will explore the effect of arteriole radius on glomerular capillary pressure and filtration in a single nephron. You can apply the concepts you learn by studying a single nephron to understand the function of the kidney as a whole.

EQUIPMENT USED The following equipment will be depicted on-screen: source beaker for blood (first beaker on left side of screen)—simulates blood flow and pressure (mm Hg) from general circulation to the nephron; drain beaker for blood (second beaker on left side of screen)—simulates the renal vein; flow tube with adjustable radius—simulates the afferent arteriole and connects the blood supply to the glomerular capillaries; second flow tube with adjustable radius—simulates the efferent arteriole and drains the glomerular capillaries into the peritubular capillaries, which ultimately drain into the renal vein (drain beaker); simulated nephron (The filtrate forms in Bowman's capsule, flows through the renal tubule—the tubular components—and empties into a collecting duct, which in turn drains into the urinary bladder.); nephron tank; glomerulus—"ball" of capillaries that forms part of the filtration membrane; glomerular (Bowman's) capsule—forms part of the filtration membrane and a capsular space where the filtrate initially forms; proximal convoluted tubule; loop of Henle; distal convoluted tubule; collecting duct; drain beaker for filtrate (beaker on right side of screen)—simulates the urinary bladder.

Experiment Instructions

Go to the home page in the PhysioEx software and click **Exercise 9: Renal System Physiology.** Click **Activity 1: The Effect of Arteriole Radius on Glomerular Filtration,** and take the online **Pre-lab Quiz** for Activity 1.

After you take the online Pre-lab Quiz, click the **Experiment** tab and begin the experiment. The experiment instructions are reprinted here for your reference. The opening screen for the experiment is shown below.

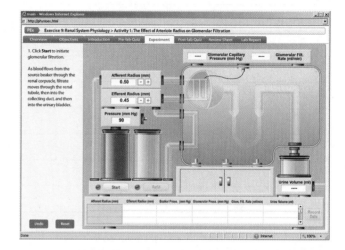

1. Click **Start** to initiate glomerular filtration. As blood flows from the source beaker through the renal corpuscle, filtrate moves through the renal tubule, then into the collecting duct, and then into the urinary bladder.

2. The glomerular capillary pressure display shows the hydrostatic blood pressure in the glomerular capillaries that promotes filtration, and the filtration rate display shows the flow rate of the fluid moving from the lumen of the glomerular capillaries into the lumen of Bowman's capsule. Click **Record Data** to display your results in the grid (and record your results in Chart 1).

3. Click **Refill** to replenish the source beaker and prepare the nephron for the next run.

> **? PREDICT Question 1**
> What will happen to the glomerular capillary pressure and filtration rate if you decrease the radius of the afferent arteriole?

4. Decrease the radius of the afferent arteriole to 0.45 mm by clicking the – button beside the afferent radius display. Click **Start** to initiate glomerular filtration.

5. Note the glomerular capillary pressure and glomerular filtration rate displays and click **Record Data** to display your results in the grid (and record your results in Chart 1).

6. Click **Refill** to replenish the source beaker and prepare the nephron for the next run.

CHART 1	Effect of Arteriole Radius on Glomerular Filtration		
Afferent arteriole radius (mm)	Efferent arteriole radius (mm)	Glomerular capillary pressure (mm Hg)	Glomerular filtration rate (ml/min)

7. You will now observe the effect of incremental decreases in the radius of the afferent arteriole.

  • Decrease the radius of the afferent arteriole by 0.05 mm by clicking the − button beside the afferent radius display.

  • Click **Start** to initiate glomerular filtration.

  • Note the glomerular capillary pressure and glomerular filtration rate displays and click **Record Data** to display your results in the grid (and record your results in Chart 1).

  • Click **Refill** to replenish the source beaker and prepare the nephron for the next run.

  Repeat this step until you reach an afferent arteriole radius of 0.35 mm.

 PREDICT Question 2
What will happen to the glomerular capillary pressure and filtration rate if you increase the radius of the afferent arteriole?

_____

8. Increase the radius of the afferent arteriole to 0.55 mm by clicking the + button beside the afferent radius display. Click **Start** to initiate glomerular filtration.

9. Note the glomerular capillary pressure and glomerular filtration rate displays and click **Record Data** to display your results in the grid (and record your results in Chart 1).

10. Click **Refill** to replenish the source beaker and prepare the nephron for the next run.

11. Increase the radius of the afferent arteriole to 0.60 mm. Click **Start** to initiate glomerular filtration.

12. Note the glomerular capillary pressure and glomerular filtration rate displays and click **Record Data** to display your results in the grid (and record your results in Chart 1).

13. Click **Refill** to replenish the source beaker and prepare the nephron for the next run.

 PREDICT Question 3
What will happen to the glomerular capillary pressure and filtration rate if you decrease the radius of the efferent arteriole?

_____

14. Decrease the radius of the afferent arteriole to 0.50 mm by clicking the − button beside the afferent radius display. Click **Start** to initiate glomerular filtration.

15. Note the glomerular capillary pressure and glomerular filtration rate displays and click **Record Data** to display your results in the grid (and record your results in Chart 1).

16. Click **Refill** to replenish the source beaker and prepare the nephron for the next run.

17. You will now observe the effect of incremental decreases in the radius of the efferent arteriole.

  • Decrease the radius of the efferent arteriole by 0.05 mm by clicking the − button beside the efferent radius display.

  • Click **Start** to initiate glomerular filtration.

- Note the glomerular capillary pressure and glomerular filtration rate displays and click **Record Data** to display your results in the grid (and record your results in Chart 1).

- Click **Refill** to replenish the source beaker and prepare the nephron for the next run.

Repeat this step until you reach an efferent arteriole radius of 0.30 mm.

After you complete the experiment, take the online **Post-lab Quiz** for Activity 1.

Activity Questions

1. Activation of sympathetic nerves that innervate the kidney leads to a decreased urine production. Knowing that fact, what do you think the sympathetic nerves do to the afferent arteriole?

_____

_____

2. How is this effect of the sympathetic nervous system beneficial? Could this effect become harmful if it goes on too long?

_____

_____

The Effect of Pressure on Glomerular Filtration

OBJECTIVES

1. To understand the terms *glomerulus, glomerular capillaries, renal tubule, filtrate, Starling forces, Bowman's capsule, renal corpuscle, afferent arteriole, efferent arteriole, glomerular capillary pressure,* and *glomerular filtration rate.*

2. To understand how changes in glomerular capillary pressure affect glomerular filtration rate.

3. To understand how changes in renal tubule pressure affect glomerular filtration rate.

Introduction

Cellular metabolism produces a complex mixture of waste products that must be eliminated from the body. This excretory function is performed by a combination of organs, most importantly, the paired kidneys. Each kidney consists of approximately one million nephrons, which carry out three crucial processes: (1) glomerular filtration, (2) tubular reabsorption, and (3) tubular secretion.

Both the blood pressure in the **glomerular capillaries** and the **filtrate** pressure in the **renal tubule** can have a significant impact on the **glomerular filtration rate.** During glomerular filtration, blood enters the **glomerulus** from the **afferent arteriole. Starling forces** (hydrostatic and osmotic pressure gradients) drive protein-free fluid between the blood in the glomerular capillaries and the filtrate in **Bowman's capsule.** The glomerular filtration rate is an index of kidney function. In humans, the filtration rate ranges from 80 to 140 ml/min, so that, in 24 hours, as much as 180 liters of filtrate is produced by the glomerular capillaries. The filtrate formed is devoid of blood cells, is essentially protein free, and contains a concentration of salts and organic molecules similar to that in blood.

Approximately 20% of the blood that enters the glomerular capillaries is normally filtered into Bowman's capsule, where it is then referred to as filtrate. The unusually high hydrostatic blood pressure in the glomerular capillaries promotes this filtration. Thus, the glomerular filtration rate can be altered by changing the afferent arteriole resistance (and, therefore, the hydrostatic pressure). In this activity you will explore the effect of blood pressure on the glomerular filtration rate in a single nephron. You can apply the concepts you learn by studying a single nephron to understand the function of the kidney as a whole.

> **EQUIPMENT USED** The following equipment will be depicted on-screen: left source beaker (first beaker on left side of screen)—simulates blood flow and pressure (mm Hg) from general circulation to the nephron; drain beaker for blood (second beaker on left side of screen)—simulates the renal vein; flow tube with adjustable radius—simulates the afferent arteriole and connects the blood supply to the glomerular capillaries; second flow tube with adjustable radius—simulates the efferent arteriole and drains the glomerular capillaries into the peritubular capillaries, which ultimately drain into the renal vein (drain beaker); simulated nephron (The filtrate forms in Bowman's capsule, flows through the renal tubule—the tubular components—and empties into a collecting duct, which in turn drains into the urinary bladder.); nephron tank; glomerulus—"ball" of capillaries that forms part of the filtration membrane; glomerular (Bowman's) capsule—forms part of the filtration membrane and a capsular space where the filtrate initially forms; proximal convoluted tubule; loop of Henle; distal convoluted tubule; collecting duct; one-way valve between end of collecting tube (duct) and urinary bladder—used to restrict the flow of filtrate into the urinary bladder, increasing the volume and pressure in the renal tubule; drain beaker for filtrate (beaker on right side of screen)—simulates the urinary bladder.

Experiment Instructions

Go to the home page in the PhysioEx software and click **Exercise 9: Renal System Physiology.** Click **Activity 2: The Effect of Pressure on Glomerular Filtration,** and take the online **Pre-lab Quiz** for Activity 2.

After you take the online Pre-lab Quiz, click the **Experiment** tab and begin the experiment. The experiment instructions are reprinted here for your reference. The opening screen for the experiment is shown on the following page.

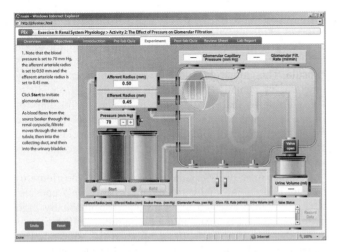

4. Increase the blood pressure to 80 mm Hg by clicking the **+** button beside the pressure display. Click **Start** to initiate glomerular filtration.

5. Note the glomerular capillary pressure and glomerular filtration rate displays and click **Record Data** to display your results in the grid (and record your results in Chart 2).

6. Click **Refill** to replenish the source beaker and prepare the nephron for the next run.

7. You will now observe the effect of further incremental increases in blood pressure.

- Increase the blood pressure by 10 mm Hg by clicking the **+** button beside the pressure display.

- Click **Start** to initiate glomerular filtration.

- Note the glomerular capillary pressure and glomerular filtration rate displays and click **Record Data** to display your results in the grid (and record your results in Chart 2).

- Click **Refill** to replenish the source beaker and prepare the nephron for the next run.

Repeat this step until you reach a blood pressure of 100 mm Hg.

1. Note that the blood pressure is set to 70 mm Hg, the afferent arteriole radius is set to 0.50 mm, and the efferent arteriole radius is set to 0.45 mm. Click **Start** to initiate glomerular filtration. As blood flows from the source beaker through the renal corpuscle, filtrate moves through the renal tubule, then into the collecting duct, and then into the urinary bladder.

2. The glomerular capillary pressure display shows the hydrostatic blood pressure in the glomerular capillaries that promotes filtration, and the filtration rate display shows the flow rate of the fluid moving from the lumen of the glomerular capillaries into the lumen of Bowman's capsule. Click **Record Data** to display your results in the grid (and record your results in Chart 2.)

3. Click **Refill** to replenish the source beaker and prepare the nephron for the next run.

? PREDICT **Question 1**
What will happen to the glomerular capillary pressure and filtration rate if you increase the blood pressure in the left source beaker?

? PREDICT **Question 2**
What will happen to the filtrate pressure in Bowman's capsule (not directly measured in this experiment) and the filtration rate if you close the one-way valve between the collecting duct and the urinary bladder?

8. Note that the valve between the collecting duct and the urinary bladder is open. Decrease the blood pressure to 70 mm Hg by clicking the button beside the pressure display. Click **Start** to initiate glomerular filtration.

CHART 2	Effect of Arteriole Radius on Glomerular Filtration			
Blood pressure (mm Hg)	Valve (open or closed)	Glomerular capillary pressure (mm Hg)	Glomerular filtration rate (ml/min)	Urine volume (ml)

9. Note the glomerular capillary pressure and glomerular filtration rate displays and click **Record Data** to display your results in the grid (and record your results in Chart 2).

10. Click **Refill** to replenish the source beaker and prepare the nephron for the next run.

11. Click the valve between the collecting duct and the urinary bladder to close it. Click **Start** to initiate glomerular filtration.

12. Note the glomerular capillary pressure and glomerular filtration rate displays and click **Record Data** to display your results in the grid (and record your results in Chart 2).

13. Click **Refill** to replenish the source beaker and prepare the nephron for the next run.

14. Increase the blood pressure to 100 mm Hg. Click **Start** to initiate glomerular filtration.

15. Note the glomerular capillary pressure and glomerular filtration rate displays and click **Record Data** to display your results in the grid (and record your results in Chart 2).

16. Click **Refill** to replenish the source beaker and prepare the nephron for the next run.

17. Click the valve between the collecting duct and the urinary bladder to open it. Click **Start** to initiate glomerular filtration.

18. Note the glomerular capillary pressure and glomerular filtration rate displays and click **Record Data** to display your results in the grid (and record your results in Chart 2).

After you complete the experiment, take the online **Post-lab Quiz** for Activity 2.

Activity Questions

1. Judging from the results in this laboratory activity, what *should be* the effect of blood pressure on glomerular filtration?

_____

_____

2. Persistent high blood pressure with inadequate glomerular filtration is now a frequent problem in Western cultures. Using the concepts in this activity, explain this health problem.

_____

_____

Renal Response to Altered Blood Pressure

OBJECTIVES

1. To understand the terms *nephron, renal tubule, filtrate, Bowman's capsule, blood pressure, afferent arteriole,* *efferent arteriole, glomerulus, glomerular filtration rate,* and *glomerular capillary pressure.*

2. To understand how blood pressure affects glomerular capillary pressure and glomerular filtration.

3. To observe which is more effective: changes in afferent or efferent arteriole radius when changes in blood pressure occur.

Introduction

In humans approximately 180 liters of filtrate flows into the **renal tubules** every day. As demonstrated in Activity 2, the **blood pressure** supplying the **nephron** can have a substantial impact on the **glomerular capillary pressure** and **glomerular filtration.** However, under most circumstances, glomerular capillary pressure and glomerular filtration remain relatively constant despite changes in blood pressure because the nephron has the capacity to alter its **afferent** and **efferent arteriole** radii.

During glomerular filtration, blood enters the **glomerulus** from the afferent arteriole. **Starling forces** (primarily hydrostatic pressure gradients) drive protein-free fluid out of the glomerular capillaries and into **Bowman's capsule.** Importantly for our body's homeostasis, a relatively constant glomerular filtration rate of 125 ml/min is maintained despite a wide range of blood pressures that occur throughout the day for an average human.

Activities 1 and 2 explored the independent effects of arteriole radii and blood pressure on glomerular capillary pressure and glomerular filtration. In the human body, these effects occur simultaneously. Therefore, in this activity, you will alter both variables to explore their combined effects on glomerular filtration and observe how changes in one variable can compensate for changes in the other to maintain an adequate glomerular filtration rate.

EQUIPMENT USED The following equipment will be depicted on-screen: left source beaker (first beaker on left side of screen)—simulates blood flow and pressure (mm Hg) from general circulation to the nephron; drain beaker for blood (second beaker on left side of screen)—simulates the renal vein; flow tube with adjustable radius—simulates the afferent arteriole and connects the blood supply to the glomerular capillaries; second flow tube with adjustable radius—simulates the efferent arteriole and drains the glomerular capillaries into the peritubular capillaries, which ultimately drain into the renal vein (drain beaker); simulated nephron (The filtrate forms in Bowman's capsule, flows through the renal tubule—the tubular components—and empties into a collecting duct, which in turn drains into the urinary bladder.); nephron tank; glomerulus—"ball" of capillaries that forms part of the filtration membrane; glomerular (Bowman's) capsule—forms part of the filtration membrane and a capsular space where the filtrate initially forms; proximal convoluted tubule; loop of Henle; distal convoluted tubule; collecting duct; one-way valve between end of collecting tube (duct) and urinary bladder—used to restrict the flow of filtrate into the urinary bladder, increasing the volume and pressure in the renal tubule; drain beaker for filtrate (beaker on right side of screen)—simulates the urinary bladder.

Experiment Instructions

Go to the home page in the PhysioEx software and click **Exercise 9: Renal System Physiology.** Click **Activity 3: Renal Response to Altered Blood Pressure,** and take the online **Pre-lab Quiz** for Activity 3.

After you take the online Pre-lab Quiz, click the **Experiment** tab and begin the experiment. The experiment instructions are reprinted here for your reference. The opening screen for the experiment is shown below.

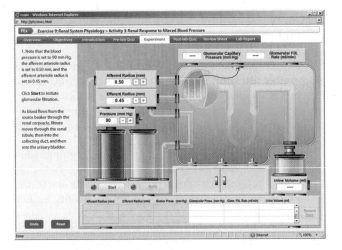

1. Note that the blood pressure is set to 90 mm Hg, the afferent arteriole radius is set to 0.50 mm, and the efferent arteriole radius is set to 0.45 mm. Click **Start** to initiate glomerular filtration. As blood flows from the source beaker through the renal corpuscle, filtrate moves through the renal tubule, then into the collecting duct, and then into the urinary bladder.

2. The glomerular capillary pressure display shows the hydrostatic blood pressure in the glomerular capillaries that promotes filtration, and the filtration rate display shows the flow rate of the fluid moving from the lumen of the glomerular capillaries into the lumen of Bowman's capsule. Click **Record Data** to display your results in the grid (and record your results in Chart 3).

3. Click **Refill** to replenish the source beaker and prepare the nephron for the next run.

4. You will now observe how the nephron might operate to keep the glomerular filtration rate relatively constant despite a large drop in blood pressure. Decrease the blood pressure to 70 mm Hg by clicking the − button beside the pressure display. Click **Start** to initiate glomerular filtration.

5. Note the glomerular capillary pressure and glomerular filtration rate displays and click **Record Data** to display your results in the grid (and record your results in Chart 3).

6. Click **Refill** to replenish the source beaker and prepare the nephron for the next run.

7. Increase the afferent arteriole radius to 0.60 mm by clicking the + button beside the afferent radius display. Click **Start** to initiate glomerular filtration.

8. Note the glomerular capillary pressure and glomerular filtration rate displays and click **Record Data** to display your results in the grid (and record your results in Chart 3).

9. Click **Refill** to replenish the source beaker and prepare the nephron for the next run.

10. Return the afferent arteriole radius to 0.50 mm by clicking the − button beside the afferent radius display and decrease the efferent radius to 0.35 mm by clicking the button beside the efferent radius display. Click **Start** to initiate glomerular filtration.

11. Note the glomerular capillary pressure and glomerular filtration rate displays and click **Record Data** to display your results in the grid (and record your results in Chart 3).

12. Click **Refill** to replenish the source beaker and prepare the nephron for the next run.

> **? PREDICT Question 1**
> What will happen to the glomerular capillary pressure and glomerular filtration rate if both of these arteriole radii changes are implemented simultaneously with the low blood pressure condition?

CHART 3	Renal Response to Altered Blood Pressure			
Afferent arteriole radius (mm)	Efferent arteriole radius (mm)	Blood pressure (mm Hg)	Glomerular capillary pressure (mm Hg)	Glomerular filtration rate (ml/min)

13. Set the afferent arteriole radius to 0.60 mm and keep the efferent arteriole radius at 0.35 mm. Click **Start** to initiate glomerular filtration.

14. Note the glomerular capillary pressure and glomerular filtration rate displays and click **Record Data** to display your results in the grid (and record your results in Chart 3).

After you complete the experiment, take the online **Post-lab Quiz** for Activity 3.

Activity Questions

1. How could an increased urine volume be viewed as beneficial to the body?

_____

_____

2. Diuretics are frequently given to people with persistent high blood pressure. Why?

_____

_____

Solute Gradients and Their Impact on Urine Concentration

OBJECTIVES

1. To understand the terms *antidiuretic hormone (ADH)*, *reabsorption, loop of Henle, collecting duct, tubule lumen, interstitial space,* and *peritubular capillaries.*

2. To explain the process of water reabsorption in specific regions of the nephron.

3. To understand the role of ADH in water reabsorption by the nephron.

4. To describe how the kidneys can produce urine that is four times more concentrated than the blood.

Introduction

As filtrate moves through the tubules of a nephron, solutes and water move *from* the **tubule lumen** *into* the **interstitial spaces** of the nephron. This movement of solutes and water relies on the total solute concentration gradient in the interstitial spaces surrounding the tubule lumen. The interstitial fluid is comprised mostly of NaCl and urea. When the nephron is permeable to solutes or water, equilibrium will be reached between the interstitial fluid and the tubular fluid contents.

Antidiuretic hormone (ADH) increases the water permeability of the **collecting duct,** allowing water to flow to areas of higher solute concentration, from the tubule lumen into the surrounding interstitial spaces. **Reabsorption** describes this movement of filtered solutes and water from the lumen of the renal tubules back into the plasma. The reabsorbed solutes and water that move into the interstitial space need to be returned to the blood, or the kidneys will rapidly swell like

balloons. The **peritubular capillaries** surrounding the renal tubule reclaim the reabsorbed substances and return them to general circulation. Peritubular capillaries arise from the efferent arteriole exiting the glomerulus and empty into the renal veins leaving the kidney.

Without reabsorption, we would excrete the solutes and water that our bodies need to maintain homeostasis. In this activity you will examine the process of passive reabsorption that occurs while filtrate travels through a nephron and urine is formed. While completing the experiment, assume that when ADH is present, the conditions favor the formation of the most concentrated urine possible.

EQUIPMENT USED The following equipment will be depicted on-screen: simulated nephron surrounded by interstitial space between the nephron and peritubular capillaries (Reabsorbed solutes, such as glucose, will move from the lumen of the tubule into the interstitial space, and then into the peritubular capillaries that branch out from the efferent arteriole.); drain beaker for filtrate—simulates the urinary bladder; antidiuretic hormone (ADH).

Experiment Instructions

Go to the home page in the PhysioEx software and click **Exercise 9: Renal System Physiology.** Click **Activity 4: Solute Gradients and Their Impact on Urine Concentration,** and take the online **Pre-lab Quiz** for Activity 4.

After you take the online Pre-lab Quiz, click the **Experiment** tab and begin the experiment. The experiment instructions are reprinted here for your reference. The opening screen for the experiment is shown below.

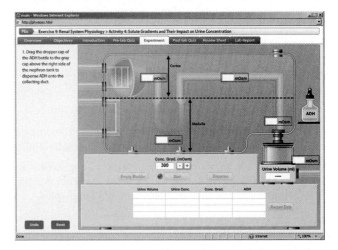

1. Drag the dropper cap of the ADH bottle to the gray cap above the right side of the nephron tank to dispense ADH onto the collecting duct.

2. Click **Dispense** beneath the concentration gradient display to adjust the maximum total solute concentration in the interstitial fluid to 300 mOsm. Because the blood solute concentration is also 300 mOsm, there is no osmotic difference between the lumen of the tubule and the surrounding interstitial fluid.

3. Click **Start** to initiate filtration. Filtrate will flow through the nephron, and solutes and water will move out of the tubules into the interstitial space. Fluid will also move back

into the peritubular capillaries, thus completing the process of reabsorption.

4. Click **Record Data** to display your results in the grid (and record your results in Chart 4).

CHART 4	Solute Gradients and Their Impact on Urine Concentration	
Urine volume (ml)	Urine concentration (mOsm)	Concentration gradient (mOsm)

5. Click **Empty Bladder** to prepare for the next run.

? PREDICT Question 1
What will happen to the urine volume and concentration as the solute gradient in the interstitial space is increased?

6. Increase the maximum concentration of the solutes in the interstitial space to 600 mOsm by clicking the **+** button beside the concentration gradient display. Click **Dispense** to adjust the maximum total solute concentration in the interstitial fluid.

7. Click **Start** to initiate filtration.

8. Click **Record Data** to display your results in the grid (and record your results in Chart 4).

9. Click **Empty Bladder** to prepare for the next run.

10. You will now observe the effect of incremental increases in maximum total solute concentration in the interstitial fluid.

- Increase the maximum concentration of the solutes in the interstitial space by 300 mOsm by clicking the **+** button beside the concentration gradient display.
- Click **Dispense** to adjust the maximum total solute concentration in the interstitial fluid.
- Click **Start** to initiate filtration.
- Click **Record Data** to display your results in the grid (and record your results in Chart 4).
- Click **Empty Bladder** to prepare for the next run.

Repeat this step until you reach the maximum total solute concentration in the interstitial fluid of 1200 mOsm.

After you complete the experiment, take the online **Post-lab Quiz** for Activity 4.

Activity Questions

1. From what you learned in this activity, speculate on ways that desert rats are able to concentrate their urine significantly more than humans.

_____

_____

2. Judging from this activity, what would be a reasonable mechanism for diuretics?

_____

_____

ACTIVITY 5

Reabsorption of Glucose via Carrier Proteins

OBJECTIVES

1. To understand the terms *reabsorption, carrier proteins, apical membrane, secondary active transport, facilitated diffusion,* and *basolateral membrane.*
2. To understand the role that glucose carrier proteins play in removing glucose from the filtrate.
3. To understand the concept of a glucose carrier transport maximum and why glucose is not normally present in the urine.

Introduction

Reabsorption is the movement of filtered solutes and water from the lumen of the renal tubules back into the plasma. Without reabsorption, we would excrete the solutes and water that our bodies require for homeostasis.

Glucose is not very large and is therefore easily filtered out of the plasma into Bowman's capsule as part of the filtrate. To ensure that glucose is reabsorbed into the body so that it can fuel cellular metabolism, glucose **carrier proteins** are present in the proximal tubule cells of the nephron. There are a finite number of these glucose carriers in each renal tubule cell. Therefore, if too much glucose is present in the filtrate, it will not all be reabsorbed and glucose will be inappropriately excreted into the urine.

Glucose is first absorbed by **secondary active transport** at the **apical membrane** of proximal tubule cells and then it leaves the tubule cell via **facilitated diffusion** along the **basolateral membrane.** Both types of carrier proteins that transport these molecules across the tubule membranes are transmembrane proteins. Because carrier proteins are needed to move glucose from the lumen of the nephron into the interstitial spaces, there is a limit to the amount of glucose that can be reabsorbed. When all glucose carriers are bound with the glucose they are transporting, excess glucose in the filtrate is eliminated in urine.

In this activity, you will examine the effect of varying the number of glucose transport proteins in the *proximal convoluted tubule*. It is important to note that, normally, the number of glucose carriers is constant in a human kidney and that it is the plasma glucose that varies during the day. Plasma glucose will be held constant in this activity, and the number of glucose carriers will be varied.

> **EQUIPMENT USED** The following equipment will be depicted on-screen: simulated nephron surrounded by interstitial space between the nephron and peritubular capillaries (Reabsorbed solutes, such as glucose, will move from the lumen of the tubule into the interstitial space, and then into the peritubular capillaries that branch out from the efferent arteriole.); drain beaker for filtrate—simulates the urinary bladder; glucose carrier protein control box—used to adjust the number of glucose carriers that will be inserted into the proximal tubule.

Experiment Instructions

Go to the home page in the PhysioEx software and click **Exercise 9: Renal System Physiology.** Click **Activity 5: Reabsorption of Glucose via Carrier Proteins,** and take the online **Pre-lab Quiz** for Activity 5.

   After you take the online Pre-lab Quiz, click the **Experiment** tab and begin the experiment. The experiment instructions are reprinted here for your reference. The opening screen for the experiment is shown below.

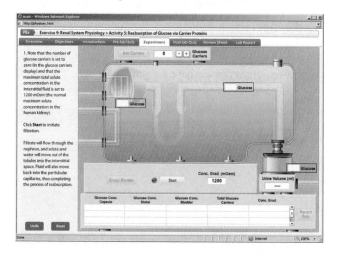

1. Note that the number of glucose carriers is set to zero (in the glucose carriers display) and that the maximum total solute concentration in the interstitial fluid is set to 1200 mOsm (the normal maximum solute concentration in the human kidney). Click **Start** to initiate filtration. Filtrate will flow through the nephron, and solute and water will move out of the tubules into the interstitial space. Fluid will also move back into the peritubular capillaries, thus completing the process of reabsorption.

2. Click **Record Data** to display your results in the grid (and record your results in Chart 5). The concentrations of glucose in Bowman's capsule, the distal convoluted tubule, and the urinary bladder will be displayed in the grid.

CHART 5	Reabsorption of Glucose via Carrier Proteins		
Glucose concentration (m*M*)			
Bowman's capsule	Distal convoluted tubule	Urinary bladder	Glucose carriers

3. Click **Empty Bladder** to prepare the nephron for the next run.

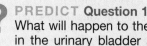

PREDICT Question 1
What will happen to the glucose concentration in the urinary bladder as glucose carriers are added to the proximal tubule?

4. Increase the number of glucose carriers to 100 (an arbitrary number) by clicking the **+** button beside the glucose carriers display. Click **Add Carriers** to insert the specified number of glucose carrier proteins per unit area into the membrane of the proximal tubule.

5. Click **Start** to initiate filtration.

6. Click **Record Data** to display your results in the grid (and record your results in Chart 5).

7. Click **Empty Bladder** to prepare the nephron for the next run.

8. You will now observe the effect of incremental increases in the number of glucose carriers.

- Increase the number of glucose carriers by 100 by clicking the **+** button beside the glucose carriers display.

- Click **Add Carriers** to insert the specified number of glucose carrier proteins per unit area into the membrane of the proximal tubule.

- Click **Start** to initiate filtration.

- Click **Record Data** to display your results in the grid (and record your results in Chart 5).

- Click **Empty Bladder** to prepare the nephron for the next run.

   Repeat this step until you have inserted 400 glucose carrier proteins per unit area into the membrane of the proximal tubule.

After you complete the experiment, take the online **Post-lab Quiz** for Activity 5.

Activity Questions

1. Why would your family physician at the turn of the twentieth century taste your urine?

_____

_____ ▬

The Effect of Hormones on Urine Formation

OBJECTIVES

1. To understand the terms *antidiuretic hormone (ADH), aldosterone, reabsorption, loop of Henle, distal convoluted tubule, collecting duct, tubule lumen,* and *interstitial space.*

2. To understand how the hormones aldosterone and ADH affect renal processes in a human kidney.

3. To understand the role of ADH in water reabsorption by the nephron.

4. To understand the role of aldosterone in solute reabsorption and secretion by the nephron.

Introduction

The concentration and volume of urine excreted by our kidneys will change depending on what our body needs for homeostasis. For example, if a person consumes a large quantity of water, the excess water will be eliminated as a large volume of dilute urine. On the other hand, when dehydration occurs, there is a clear benefit in being able to produce a small volume of concentrated urine to retain water. Activity 4 demonstrated how the total solute concentration gradient in the interstitial spaces surrounding the tubule lumen makes it possible to excrete concentrated urine.

Aldosterone is a hormone produced by the adrenal cortex under the control of the body's *renin-angiotensin system.* A decrease in blood pressure is detected by cells in the afferent arteriole, triggering the release of renin. Renin acts as a proteolytic enzyme, causing angiotensinogen to be converted into angiotensin I. Endothelial cells throughout the body possess a *converting enzyme* that converts angiotensin I into angiotensin II. Angiotensin II signals the adrenal cortex to secrete aldosterone. Aldosterone acts on the distal convoluted tubule cells in the nephron to promote the reabsorption of sodium from filtrate *into* the body and the secretion of potassium *from* the body. This electrolyte shift, coupled with the addition of **antidiuretic hormone (ADH),** also causes more water to be reabsorbed into the blood, resulting in increased blood pressure.

ADH is manufactured by the hypothalamus and stored in the posterior pituitary gland. ADH levels are influenced by the osmolality of body fluids and the volume and pressure of the cardiovascular system. A 1% change in body osmolality will cause this hormone to be secreted. The primary action of this hormone is to increase the permeability of the collecting duct to water so that more water is reabsorbed into the body

by inserting aquaporins, or water channels, in the apical membrane. Without this water reabsorption, the body would quickly dehydrate.

Thus, our kidneys tightly regulate the amount of water and solutes excreted to maintain water balance in the body. If water intake is down, or if there has been a fluid loss from the body, the kidneys work to conserve water by making the urine very hyperosmotic (having a relatively high solute concentration) to the blood. If there has been a large intake of fluid, the urine is more hypo-osmotic. In the normal individual, urine osmolarity varies from 50 to 1200 milliosmoles/kg of water.

EQUIPMENT USED The following equipment will be depicted on-screen: simulated nephron surrounded by interstitial space between the nephron and peritubular capillaries (Reabsorbed solutes, such as glucose, will move from the lumen of the tubule into the interstitial space, and then into the peritubular capillaries that branch out from the efferent arteriole.); drain beaker for filtrate—simulates the urinary bladder; aldosterone; antidiuretic hormone (ADH).

Experiment Instructions

Go to the home page in the PhysioEx software and click **Exercise 9: Renal System Physiology.** Click **Activity 6: The Effect of Hormones on Urine Formation,** and take the online **Pre-lab Quiz** for Activity 6.

After you take the online Pre-lab Quiz, click the **Experiment** tab and begin the experiment. The experiment instructions are reprinted here for your reference. The opening screen for the experiment is shown below.

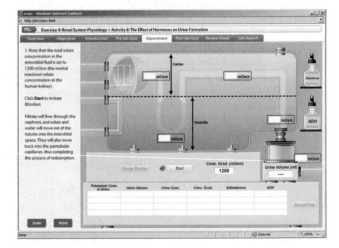

1. Note that the total solute concentration in the interstitial fluid is set to 1200 mOsm (the normal maximum solute concentration in the human kidney). Click **Start** to initiate filtration. Filtrate will flow through the nephron, and solute and water will move out of the tubules into the interstitial space. They will also move back into the peritubular capillaries, thus completing the process of reabsorption.

2. Click **Record Data** to display your results in the grid (and record your results in Chart 6). You will use this baseline data to compare the conditions of the filtrate and urine volume in the presence of the hormones aldosterone and ADH.

CHART 6	The Effect of Hormones on Urine Formation			
Potassium concentration (m*M*)	Urine volume (ml)	Urine concentration (mOsm)	Aldosterone	ADH

3. Click **Empty Bladder** to prepare the nephron for the next run.

PREDICT Question 1
What will happen to the urine volume (compared with baseline) when aldosterone is added to the distal tubule?

4. Drag the dropper cap of the aldosterone bottle to the gray cap above the right side of the nephron tank to dispense aldosterone into the tank surrounding the distal tubule and the collecting duct.

5. Click **Start** to initiate filtration.

6. Click **Record Data** to display your results in the grid (and record your results in Chart 6).

7. Click **Empty Bladder** to prepare the nephron for the next run.

8. Drag the dropper cap of the ADH bottle to the gray cap at the top right side of the nephron tank to dispense ADH into the tank surrounding the distal tubule and the collecting duct.

PREDICT Question 2
What will happen to the urine volume (compared with baseline) when ADH is added to the collecting duct?

9. Click **Start** to initiate filtration.

10. Click **Record Data** to display your results in the grid (and record your results in Chart 6).

11. Click **Empty Bladder** to prepare the nephron for the next run.

PREDICT Question 3
What will happen to the urine volume and the urine concentration (compared with baseline) in the presence of both aldosterone and ADH?

12. Drag the dropper cap of the aldosterone bottle and then the dropper cap of the ADH bottle to the gray cap above the right side of the nephron tank to dispense aldosterone and ADH into the tank surrounding the distal convoluted tubule and the collecting duct.

13. Click **Start** to initiate filtration.

14. Click **Record Data** to display your results in the grid (and record your results in Chart 6).

After you complete the experiment, take the online **Post-lab Quiz** for Activity 6.

Activity Questions

1. Why does ethanol consumption lead to a dramatic increase in urine production?

_____

_____

2. Why do angiotensin converting enzyme (ACE) inhibitors given to people with hypertension lead to increased urine production?

_____

_____

NAME _____

LAB TIME/DATE _____

Renal System Physiology

A C T I V I T Y 1 The Effect of Arteriole Radius on Glomerular Filtration

1. What are two primary functions of the kidney? _____

   _____

2. What are the components of the renal corpuscle? _____

   _____

3. Starting at the renal corpuscle, list the components of the renal tubule as they are encountered by filtrate. _____

   _____

   _____

4. Describe the effect of decreasing the afferent arteriole radius on glomerular capillary pressure and filtration rate. How well

   did the results compare with your prediction? _____

   _____

5. Describe the effect of increasing the afferent arteriole radius on glomerular capillary pressure and filtration rate. How well

   did the results compare with your prediction? _____

   _____

6. Describe the effect of decreasing the efferent arteriole radius on glomerular capillary pressure and filtration rate. How well

   did the results compare with your prediction? _____

   _____

7. Describe the effect of increasing the efferent radius on glomerular capillary pressure and filtration rate._____

   _____

A C T I V I T Y 2 The Effect of Pressure on Glomerular Filtration

1. As blood pressure increased, what happened to the glomerular capillary pressure and the glomerular filtration rate? How well

   did the results compare with your prediction? _____

   _____

2. Compare the urine volume in your baseline data with the urine volume as you increased the blood pressure. How did the

   urine volume change? _____

   _____

3. How could the change in urine volume with the increase in blood pressure be viewed as being beneficial to the body?

   _____

   _____

4. When the one-way valve between the collecting duct and the urinary bladder was closed, what happened to the filtrate pressure
   in Bowman's capsule (this is not directly measured in this experiment) and the glomerular filtration rate? How well did

   the results compare with your prediction? _____

   _____

5. How did increasing the blood pressure alter the results when the valve was closed? _____

   _____

ACTIVITY 3   Renal Response to Altered Blood Pressure

1. List the several mechanisms you have explored that change the glomerular filtration rate. How does each mechanism specifically

   alter the glomerular filtration rate? _____

   _____

2. Describe and explain what happened to the glomerular capillary pressure and glomerular filtration rate when *both* arteriole
   radii changes were implemented simultaneously with the low blood pressure condition. How well did the results compare

   with your prediction? _____

   _____

   _____

3. How could you adjust the afferent or efferent radius to compensate for the effect of reduced blood pressure on the glomerular

   filtration rate? _____

   _____

4. Which arteriole radius adjustment was more effective at compensating for the effect of low blood pressure on the glomerular

   filtration rate? Explain why you think this difference occurs. _____

   _____

5. In the body, how does a nephron maintain a near-constant glomerular filtration rate despite a constantly fluctuating blood

   pressure? _____

   _____

ACTIVITY 4   Solute Gradients and Their Impact on Urine Concentration

1. What happened to the urine concentration as the solute concentration in the interstitial space was increased? How well did the results compare to your prediction? _____

_____

_____

2. What happened to the volume of urine as the solute concentration in the interstitial space was increased? How well did the results compare to your prediction? _____

_____

3. What do you think would happen to urine volume if you did not add ADH to the collecting duct? _____

_____

_____

4. Is most of the tubule filtrate reabsorbed into the body or excreted in urine? Explain. _____

_____

_____

5. Can the reabsorption of solutes influence water reabsorption from the tubule fluid? Explain. _____

_____

ACTIVITY 5   Reabsorption of Glucose via Carrier Proteins

1. What happens to the concentration of glucose in the urinary bladder as the number of glucose carriers increases? _____

_____

_____

2. What types of transport are utilized during glucose reabsorption and where do they occur? _____

_____

3. Why does the glucose concentration in the urinary bladder become zero in these experiments? _____

_____

4. A person with type 1 diabetes cannot make insulin in the pancreas, and a person with untreated type 2 diabetes does not respond to the insulin that is made in the pancreas. In either case, why would you expect to find glucose in the person's urine?

_____

_____

ACTIVITY 6 The Effect of Hormones on Urine Formation

1. How did the addition of aldosterone affect urine volume (compared with baseline)? Can the reabsorption of solutes influence

   water reabsorption in the nephron? Explain. How well did the results compare with your prediction? _____

   _____

   _____

2. How did the addition of ADH affect urine volume (compared with baseline)? How well did the results compare with your
   prediction? Why did the addition of ADH also affect the concentration of potassium in the urine (compared with baseline)?

   _____

   _____

3. What is the principal determinant for the release of aldosterone from the adrenal cortex? _____

   _____

4. How did the addition of both aldosterone and ADH affect urine volume (compared with baseline)? How well did the results

   compare with your prediction? _____

   _____

   _____

5. What is the principal determinant for the release of ADH from the posterior pituitary gland? Does ADH favor the formation

   of dilute or concentrated urine? Explain why. _____

   _____

   _____

6. Which hormone (aldosterone or ADH) has the greater effect on urine volume? Why? _____

   _____

   _____

7. If ADH is not available, can the urine concentration still vary? Explain your answer. _____

   _____

   _____

8. Consider this situation: you want to reabsorb sodium ions but you do not want to increase the volume of the blood by reabsorbing
   large amounts of water from the filtrate. Assuming that aldosterone and ADH are both present, how would you adjust

   the hormones to accomplish the task? _____

   _____

Acid-Base Balance

Exercise Overview

pH denotes the hydrogen ion concentration, $[H^+]$, in a solution (such as body fluids). The reciprocal relationship between pH and $[H^+]$ is defined by the following equation.

$$pH = \log(1/[H^+])$$

Because the relationship is reciprocal, $[H^+]$ is higher at *lower* pH values (indicating higher acid levels) and lower at *higher* pH values (indicating lower acid levels).

The pH of a body's fluid is also referred to as its **acid-base balance.** An **acid** is a substance that releases H^+ in solution. A **base,** often a hydroxyl ion (OH^-) or bicarbonate ion (HCO_3^-), is a substance that binds, or buffers, the H^+. A **strong acid** completely dissociates in solution, releasing all of its hydrogen ions and, thus, lowering the solution's pH. A **weak acid** dissociates incompletely and does not release all of its hydrogen ions in solution, producing a lesser effect on the solution's pH. A **strong base** has a strong tendency to bind to H^+, raising the solution's pH. A **weak base** binds less of the H^+, producing a lesser effect on the solution's pH.

The pH of body fluids is very tightly regulated. Blood and tissue fluids normally have a pH between 7.35 and 7.45. Under pathological conditions, blood pH as low as 6.9 or as high as 7.8 has been recorded, but a higher or lower pH cannot sustain human life. The narrow range from 7.35 to 7.45 is remarkable when you consider the vast number of biochemical reactions that take place in the body. The human body normally produces a large amount of H^+ as the result of metabolic processes; ingested acids; and the products of fat, sugar, and amino acid metabolism. The regulation of a relatively constant internal pH is one of the major physiological functions of the body's organ systems.

To maintain pH homeostasis, the body utilizes both *chemical* and *physiological* buffering systems. Chemical buffers are composed of a mixture of weak acids and weak bases. They help regulate the body's pH levels by binding H^+ and removing it from solution as its concentration begins to rise or by releasing H^+ into solution as its concentration begins to fall. The body's three major chemical buffering systems are the *bicarbonate, phosphate,* and *protein buffer systems.* We will not focus on chemical buffering systems in this exercise, but keep in mind that chemical buffers are the fastest form of compensation and can return pH to normal within a fraction of a second.

The body's two major physiological buffering systems are the **renal system** and the **respiratory system.** The renal system is the slower of the two, taking hours to days to do its work. The respiratory system usually works within minutes, but cannot handle the amount of pH change that the renal system can. These physiological buffer systems help regulate body pH by controlling the output of acids, bases, or carbon dioxide (CO_2) from the body. For example, if there is too much acid in the body, the renal system may respond by excreting more H^+ from the body in urine. Similarly, if there is too much carbon dioxide in the blood, the respiratory system may respond by increasing ventilation to expel the excess carbon dioxide. Carbon dioxide levels have a direct effect on pH because the addition of carbon dioxide to the blood results in the generation of more H^+. The following equation shows what happens when carbon dioxide combines with water in the blood, producing carbonic acid.

$$H_2O + CO_2 \rightleftarrows \underset{\substack{\text{carbonic} \\ \text{acid}}}{H_2CO_3} \rightleftarrows H^+ + \underset{\substack{\text{bicarbonate} \\ \text{ion}}}{HCO_3^-}$$

Hyperventilation

OBJECTIVES

1. To introduce pH homeostasis in the body.
2. To understand the normal ranges for pH and P_{CO_2}.
3. To recognize respiratory alkalosis and its causes.
4. To interpret an oscilloscope tracing for hyperventilation and compare it with a tracing for normal breathing.

Introduction

Acid-base imbalances can have respiratory and metabolic causes. When diagnosing these disorders, two key signs are evaluated: the pH and the partial pressure of carbon dioxide in the blood (P_{CO_2}). The normal range for pH is between 7.35 and 7.45, and the normal range for P_{CO_2} is between 35 and 45 mm Hg. When the pH falls below 7.35, the body is said to be in a state of **acidosis.** When the pH rises above 7.45, the body is said to be in a state of **alkalosis.**

Respiratory alkalosis is the condition of too little carbon dioxide in the blood. Respiratory alkalosis commonly results from traveling to high altitude (where the air contains less oxygen) or hyperventilation, which can be brought on by fever, panic attack, or anxiety. Hyperventilation, defined as an increase in the rate and depth of breathing, removes carbon dioxide from the blood faster than it is being produced by the cells of the body, reducing the amount of H^+ in the blood and, thus, increasing the blood's pH. The following equation shows the shift in the equilibrium that results in the increase in blood pH due to less carbon dioxide in the blood.

$$H_2O + CO_2 \leftarrow \underset{\substack{\text{carbonic} \\ \text{acid}}}{H_2CO_3} \leftarrow H^+ + \underset{\substack{\text{bicarbonate} \\ \text{ion}}}{HCO_3^-}$$

The renal system can compensate for alkalosis by retaining H^+ and excreting bicarbonate ions to lower the blood pH levels back to the normal range.

EQUIPMENT USED The following equipment will be depicted on-screen: simulated lung chamber; pH meter; oscilloscope; two breathing patterns: normal and hyperventilation.

Experiment Instructions

Go to the home page in the PhysioEx software and click **Exercise 10: Acid-Base Balance.** Click **Activity 1: Hyperventilation,** and take the online **Pre-lab Quiz** for Activity 1.

After you take the online Pre-lab Quiz, click the **Experiment** tab and begin the experiment. The experiment instructions are reprinted here for your reference. The opening screen for the experiment is shown below.

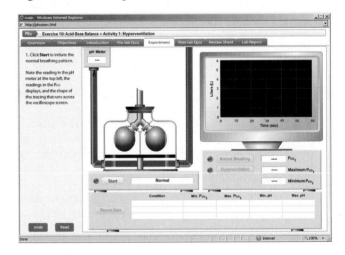

1. Click **Start** to initiate the normal breathing pattern. Note the reading in the pH meter at the top left, the readings in the P_{CO_2} displays, and the shape of the tracing that runs across the oscilloscope screen.

2. Click **Record Data** to display your results in the grid (and record your results in Chart 1).

? PREDICT Question 1
What do you think will happen to the pH and P_{CO_2} levels with hyperventilation?

CHART 1	Hyperventilation Breathing Patterns				
Condition		Minimum P_{CO_2}	Maximum P_{CO_2}	Minimum pH	Maximum pH

3. Click **Start** to initiate the normal breathing pattern. After the normal breathing tracing runs for 10 seconds, click **Hyperventilation** to initiate the hyperventilation breathing pattern. Note the reading in the pH meter at the top left, the readings in the P_{CO_2} displays, and the shape of the tracing that runs across the oscilloscope screen.

4. Click **Record Data** to display your results in the grid (and record your results in Chart 1).

5. Click **Start** to initiate the normal breathing pattern. After the normal breathing tracing runs for 10 seconds, click **Hyperventilation** to initiate the hyperventilation breathing pattern. After the hyperventilation tracing runs for 10 seconds, click **Normal Breathing** to return to the normal breathing pattern. Note the reading in the pH meter at the top left, the readings in the P_{CO_2} displays, and the shape of the tracing that runs across the oscilloscope screen.

6. Click **Record Data** to display your results in the grid (and record your results in Chart 1).

After you complete the experiment, take the online **Post-lab Quiz** for Activity 1.

Activity Questions

1. At what pH range is the body considered to be in a state of respiratory alkalosis?

_____

_____

2. How can the body compensate for respiratory alkalosis?

_____

_____

3. How did the tidal volume change with hyperventilation?

_____

_____

4. What might cause a person to hyperventilate?

_____

_____

Rebreathing

OBJECTIVES

1. To understand how rebreathing can simulate hypoventilation.
2. To observe the results of respiratory acidosis.
3. To describe the causes of respiratory acidosis.

Introduction

The body is said to be in a state of **acidosis** when the pH of the blood falls below 7.35 (although a pH of 7.35 is technically not acidic). Respiratory acidosis is the result of impaired respiration, or *hypoventilation*, which leads to the accumulation of too much carbon dioxide in the blood. The causes of impaired respiration include airway obstruction, depression of the respiratory center in the brain stem, lung disease (such as emphysema and chronic bronchitis), and drug overdose.

Recall that carbon dioxide contributes to the formation of carbonic acid when it combines with water through a reversible reaction catalyzed by carbonic anhydrase. The carbonic acid then dissociates into hydrogen ions and bicarbonate ions. Because hypoventilation results in elevated carbon dioxide levels in the blood, the equilibrium shifts, the H^+ levels increase, and the pH value of the blood decreases.

$$H_2O + CO_2 \rightarrow \underset{\substack{\text{carbonic} \\ \text{acid}}}{H_2CO_3} \rightarrow H^+ + \underset{\substack{\text{bicarbonate} \\ \text{ion}}}{HCO_3^-}$$

Rebreathing is the action of breathing in air that was just expelled from the lungs. Rebreathing results in the accumulation of carbon dioxide in the blood. Breathing into a paper bag is an example of rebreathing. (Note that breathing into a paper bag can deplete the body of oxygen and is therefore not the best therapy for hyperventilation because it can mask other life-threatening emergencies, such as a heart attack or asthma.) In this activity, you will observe what happens to pH and carbon dioxide levels in the blood during rebreathing. In the body, the kidneys regulate the acid-base balance by altering the amount of H^+ and HCO_3^- excreted in the urine.

EQUIPMENT USED The following equipment will be depicted on-screen: simulated lung chamber; pH meter; oscilloscope; two breathing patterns: normal and rebreathing.

Experiment Instructions

Go to the home page in the PhysioEx software and click **Exercise 10: Acid-Base Balance.** Click **Activity 2: Rebreathing,** and take the online **Pre-lab Quiz** for Activity 2.

After you take the online Pre-lab Quiz, click the **Experiment** tab and begin the experiment. The experiment instructions are reprinted here for your reference. The opening screen for the experiment is shown below.

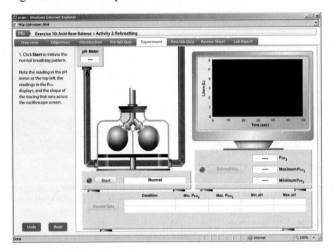

1. Click **Start** to initiate the normal breathing pattern. Note the reading in the pH meter at the top left, the readings in the P_{CO_2} displays, and the shape of the tracing that runs across the oscilloscope screen.

2. Click **Record Data** to display your results in the grid (and record your results in Chart 2).

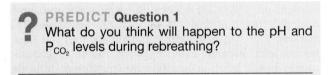

PREDICT Question 1
What do you think will happen to the pH and P_{CO_2} levels during rebreathing?

3. Click **Start** to initiate the normal breathing pattern. After the normal breathing tracing runs for 10 seconds, click **Rebreathing** to initiate the rebreathing pattern. Note the reading in the pH meter at the top left, the readings in the P_{CO_2} displays, and the shape of the tracing that runs across the oscilloscope screen.

4. Click **Record Data** to display your results in the grid (and record your results in Chart 2).

After you complete the experiment, take the online **Post-lab Quiz** for Activity 1.

Activity Questions

1. Did the pH level of the blood change at all with rebreathing? If so, how did it change?

_____

_____

2. What happens to the pH level of the blood when there is too much carbon dioxide remaining in the blood?

_____

_____

3. How did the tidal volumes change with rebreathing?

_____

_____

4. Describe two ways in which too much carbon dioxide might remain in the blood.

_____

_____

_____

ACTIVITY 3

Renal Responses to Respiratory Acidosis and Respiratory Alkalosis

OBJECTIVES

1. To understand renal compensation mechanisms for respiratory acidosis and respiratory alkalosis.

2. To explore the functional unit of the kidneys that responds to acid-base balance.

3. To observe the changes in ion concentrations that occur with renal compensation.

Introduction

The kidneys play a major role in maintaining fluid, electrolyte, and acid-base balance in the body's internal environment. By regulating the amount of water lost in the urine, the kidneys defend the body against excessive hydration or dehydration. By regulating the acidity of urine and the rate of electrolyte excretion, the kidneys maintain plasma pH and electrolyte levels within normal limits.

CHART 2	Normal Breathing Patterns			
Condition	Minimum P_{CO_2}	Maximum P_{CO_2}	Minimum pH	Maximum pH

Renal compensation is the body's primary method of compensating for conditions of respiratory acidosis or respiratory alkalosis. The kidneys regulate the acid-base balance by altering the amount of H^+ and HCO_3^- excreted in the urine. If we revisit the equation for the dissociation of carbonic acid, a weak acid, we see that the conservation of bicarbonate ion (base) has the same net effect as the loss of acid, H^+.

$$H_2O + CO_2 \rightleftarrows \underset{\substack{\text{carbonic} \\ \text{acid}}}{H_2CO_3} \rightleftarrows H^+ + \underset{\substack{\text{bicarbonate} \\ \text{ion}}}{HCO_3^-}$$

In this activity you will examine how the renal system compensates for respiratory acidosis or respiratory alkalosis. Respiratory acidosis is generally caused by the accumulation of carbon dioxide in the blood from hypoventilation, but it can also be caused by rebreathing. Acidosis results in a lower-than-normal blood pH. Respiratory alkalosis is caused by a depletion of carbon dioxide, often caused by an episode of hyperventilation, and results in an elevated blood pH.

You will primarily be working with the variable P_{CO_2}. Recall that the normal range for pH is between 7.35 and 7.45 and the normal range for P_{CO_2} is between 35 and 45 mm Hg. You will observe how increases and decreases in P_{CO_2} affect the levels of H^+ and HCO_3^- that the kidneys excrete in urine. The functional unit for adjusting the plasma composition is the **nephron.** Remember that although the renal system can partially compensate for pH imbalances with a respiratory cause, the kidneys cannot fully compensate if respirations have not returned to normal because the carbon dioxide levels will still be abnormal.

EQUIPMENT USED The following equipment will be depicted on-screen: source beaker for blood (first beaker on left side of screen); drain beaker for blood (second beaker on left side of screen); simulated nephron (The filtrate forms in Bowman's capsule and flows through the renal tubule—the tubular components, and empties into a collecting duct which, in turn drains into the urinary bladder.); nephron tank; glomerulus—"ball" of capillaries that forms part of the filtration membrane; glomerular (Bowman's) capsule—forms part of the filtration membrane and a capsular space where the filtrate initially forms; proximal convoluted tubule; loop of Henle; distal convoluted tubule; collecting duct; drain beaker for filtrate (beaker on right side of screen)—simulates the urinary bladder.

Experiment Instructions

Go to the home page in the PhysioEx software and click **Exercise 10: Acid-Base Balance.** Click **Activity 3: Renal Responses to Respiratory Acidosis and Respiratory Alkalosis** and take the online **Pre-lab Quiz** for Activity 3.

After you take the online Pre-lab Quiz, click the **Experiment** tab and begin the experiment. The experiment instructions are reprinted here for your reference. The opening screen for the experiment is shown above.

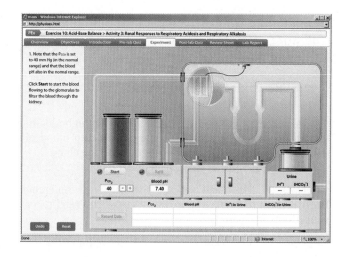

1. Note that the P_{CO_2} is set to 40 mm Hg (in the normal range) and that the blood pH is also in the normal range. Click **Start** to start the blood flowing to the glomerulus to filter the blood through the kidney.

2. Note the $[H^+]$ and $[HCO_3^-]$ in the urine and click **Record Data** to display your results in the grid (and record your results in Chart 3).

CHART 3	Renal Responses to Respiratory Acidosis and Respiratory Alkalosis		
P_{CO_2}	Blood pH	$[H^+]$ in urine	$[HCO_3^-]$ in urine

3. Click **Refill** to replenish the source beaker.

? PREDICT Question 1
What effect do you think lowering the will have on $[H^+]$ and $[HCO_3^-]$ in the urine?

4. Lower the P_{CO_2} to 30 by clicking the – button beside the P_{CO_2} display. Note the corresponding increase in blood pH (above the normal range). Click **Start** to start the blood flowing to the glomerulus to filter the blood through the kidney.

5. Note the $[H^+]$ and $[HCO_3^-]$ in the urine and click **Record Data** to display your results in the grid (and record your results in Chart 3).

6. Click **Refill** to replenish the source beaker.

? PREDICT Question 2
What effect do you think raising the P_{CO_2} will
have on [H^+] and [HCO_3^-] in the urine?

7. Raise the P_{CO_2} to 60 by clicking the **+** button beside the
P_{CO_2} display. Note the corresponding decrease in blood pH
(below the normal range). Click **Start** to start the blood flowing to the glomerulus to filter the blood through the kidney.

8. Note the [H^+] and [HCO_3^-] in the urine and click
Record Data to display your results in the grid (and record
your results in Chart 3).

After you complete the experiment, take the online **Post-lab
Quiz** for Activity 3.

Activity Questions

1. Describe how the kidneys respond to respiratory acidosis.

_____

_____

2. What P_{CO_2} corresponded to respiratory acidosis?

_____

3. Describe how the kidneys respond to respiratory alkalosis.

_____

_____

4. What P_{CO_2} corresponded to respiratory alkalosis?

_____

ACTIVITY 4

Respiratory Responses to Metabolic Acidosis and Metabolic Alkalosis

OBJECTIVES

1. To understand the causes of metabolic acidosis and
metabolic alkalosis.
2. To observe the physiological changes that occur with an
increase and decrease in metabolic rate.
3. To explain how the respiratory system compensates for
metabolic acidosis and alkalosis.

Introduction

Conditions of acidosis and alkalosis that do not have respiratory causes are termed *metabolic acidosis and metabolic
alkalosis*. **Metabolic acidosis** is characterized by low

plasma HCO_3^- and pH. The causes of metabolic acidosis
include:

- **Ketoacidosis,** a buildup of keto acids that can result
from diabetes mellitus
- **Salicylate poisoning,** a toxic condition resulting from
ingestion of too much aspirin or oil of wintergreen (a
substance often found in laboratories)
- The ingestion of too much alcohol, which metabolizes
into acetic acid
- Diarrhea, which results in the loss of bicarbonate with
the elimination of intestinal contents
- Strenuous exercise, which can cause a buildup of lactic
acid from anaerobic muscle metabolism

Metabolic alkalosis is characterized by elevated plasma
HCO_3^- and pH. The causes of metabolic alkalosis include:

- Ingestion of alkali, such as antacids or bicarbonate
- Vomiting, which can result in the loss of too much H^+
- Constipation, which may result in significant reabsorp-
tion of HCO_3^-

Increases or decreases in the body's normal metabolic rate
can also result in metabolic acidosis or alkalosis. Recall
that carbon dioxide—a waste product of metabolism—mixes
with water in plasma to form carbonic acid, which in turn
forms H^+.

$$H_2O + CO_2 \rightleftarrows \underset{\substack{\text{carbonic} \\ \text{acid}}}{H_2CO_3} \rightleftarrows H^+ + \underset{\substack{\text{bicarbonate} \\ \text{ion}}}{HCO_3^-}$$

An increase in the normal metabolic rate causes more carbon
dioxide to form as a metabolic waste product, resulting in the
formation of more H^+ and, therefore, lower plasma pH,
potentially causing acidosis. Other acids that are also normal
metabolic waste products (such as ketone bodies and phos-
phoric, uric, and lactic acids) would likewise accumulate with
an increase in metabolic rate.

Conversely, a decrease in the normal metabolic rate
causes less carbon dioxide to form as a metabolic waste product, resulting in the formation of less H^+ and, therefore,
higher plasma pH, potentially causing alkalosis. Many factors
can affect the rate of cell metabolism. For example, fever,
stress, or the ingestion of food all cause the rate of cell metabolism to *increase*. Conversely, a fall in body temperature or a
decrease in food intake causes the rate of cell metabolism to
decrease.

The respiratory system compensates for metabolic
acidosis or alkalosis by expelling or retaining carbon
dioxide in the blood. During metabolic acidosis, respiration
increases to expel carbon dioxide from the blood, thus
decreasing [H^+] and raising the pH. During metabolic
alkalosis, respiration decreases to promote the accumulation
of carbon dioxide in the blood, thus increasing [H^+] and
decreasing the pH.

The renal system also compensates for metabolic acidosis and alkalosis by conserving or excreting bicarbonate ions.
Nevertheless, in this activity, you will focus on respiratory
compensation of metabolic acidosis and alkalosis.

Experiment Instructions

Go to the home page in the PhysioEx software and click **Exercise 10: Acid-Base Balance.** Click **Activity 4: Respiratory Responses to Metabolic Acidosis and Metabolic Alkalosis,** and take the online **Pre-lab Quiz** for Activity 4.

After you take the online Pre-lab Quiz, click the **Experiment** tab and begin the experiment. The experiment instructions are reprinted here for your reference. The opening screen for the experiment is shown below.

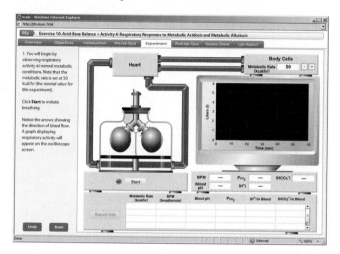

1. You will begin by observing respiratory activity at normal metabolic conditions. Note that the metabolic rate is set at 50 kcal/hr (the normal value for this experiment). Click **Start** to initiate breathing and blood flow. Notice the arrows showing the direction of blood flow. A graph displaying respiratory activity will appear on the oscilloscope screen.

2. Note the data in the displays below the oscilloscope screen and click **Record Data** to display your results in the grid (and record your results in Chart 4).

3. Increase the metabolic rate to 60 kcal/hr by clicking the + button beside the metabolic rate display. Click **Start** to initiate breathing and blood flow.

4. Note the data in the displays below the oscilloscope screen and click **Record Data** to display your results in the grid (and record your results in Chart 4).

5. Click **Clear Tracings** to clear the tracings on the oscilloscope.

? PREDICT Question 1
What do you think will happen when the metabolic rate is increased to 80 kcal/hr?

6. Increase the metabolic rate to 80 kcal/hr by clicking the + button beside the metabolic rate display. Click **Start** to initiate breathing and blood flow.

7. Note the data in the displays below the oscilloscope screen and click **Record Data** to display your results in the grid (and record your results in Chart 4).

8. Click **Clear Tracings** to clear the tracings on the oscilloscope.

9. Decrease the metabolic rate to 40 kcal/hr by clicking the − button beside the metabolic rate display. Click **Start** to initiate breathing and blood flow.

10. Note the data in the displays below the oscilloscope screen and click **Record Data** to display your results in the grid (and record your results in Chart 4).

11. Click **Clear Tracings** to clear the tracings on the oscilloscope.

? PREDICT Question 2
What do you think will happen when the metabolic rate is decreased to 20 kcal/hr?

CHART 4	Respiratory Responses to Metabolic Acidosis and Metabolic Alkalosis				
Metabolic rate	BPM (breaths/min)	Blood pH	P_{CO_2}	$[H^+]$ in blood	$[HCO_3^-]$ in blood

12. Decrease the metabolic rate to 20 kcal/hr by clicking the button beside the metabolic rate display. Click **Start** to initiate breathing and blood flow.

13. Note the data in the displays below the oscilloscope screen and click **Record Data** to display your results in the grid (and record your results in Chart 4).

After you complete the experiment, take the online **Post-lab Quiz** for Activity 4.

Activity Questions

1.   Describe what happens to carbon dioxide and pH with increased metabolism.

_____

_____

2.   Describe the respiratory response to metabolic acidosis.

_____

_____

3.   When the respiratory system compensates for the metabolic acidosis, does the pH increase or decrease in value?

_____

_____

4.   Describe the respiratory response to metabolic alkalosis.

_____

_____

NAME_____

LAB TIME/DATE_____

Acid-Base Balance

ACTIVITY 1 Hyperventilation

1. Describe the normal ranges for pH and carbon dioxide in the blood. _____

_____

_____

2. Describe what happened to the pH and the carbon dioxide levels with hyperventilation. How well did the results compare

   with your prediction?_____

_____

_____

3. Explain how returning to normal breathing after hyperventilation differed from hyperventilation without returning to normal

   breathing. _____

_____

_____

4. Describe some possible causes of respiratory alkalosis. _____

_____

_____

ACTIVITY 2 Rebreathing

1. Describe what happened to the pH and the carbon dioxide levels during rebreathing. How well did the results compare with

   your prediction?_____

_____

_____

2. Describe some possible causes of respiratory acidosis._____

_____

_____

3. Explain how the renal system would compensate for respiratory acidosis. _____

_____

_____

ACTIVITY 3 Renal Responses to Respiratory Acidosis and Respiratory Alkalosis

1. Describe what happened to the concentration of ions in the urine when the P_{CO_2} was lowered. How well did the results

   compare with your prediction? _____

   _____

   _____

2. What condition was simulated when the P_{CO_2} was lowered?_____

   _____

   _____

3. Describe what happened to the concentration of ions in the urine when the P_{CO_2} was raised. How well did the results compare

   with your prediction?_____

   _____

   _____

4. What condition was simulated when the P_{CO_2} was raised? _____

   _____

   _____

ACTIVITY 4 Respiratory Responses to Metabolic Acidosis and Metabolic Alkalosis

1. Describe what happened to the blood pH when the metabolic rate was increased to 80 kcal/hr. What body system was

   compensating? How well did the results compare with your prediction?_____

   _____

   _____

2. List and describe some possible causes of metabolic acidosis._____

   _____

   _____

3. Describe what happened to the blood pH when the metabolic rate was decreased to 20 kcal/hr. What body system was compensating? How well did the results compare with your prediction? _____

_____

_____

4. List and describe some possible causes of metabolic alkalosis. _____

_____

_____

Blood Analysis

Exercise Overview

Blood transports soluble substances to and from all cells of the body. Laboratory analysis of our blood can reveal important information about how well this function is being achieved. The five activities in this exercise simulate common laboratory tests performed on blood: (1) *hematocrit* determination, (2) *erythrocyte sedimentation rate*, (3) *hemoglobin* determination, (4) *blood typing,* and (5) total *cholesterol* determination.

Hematocrit refers to the percentage of red blood cells (RBCs), or erythrocytes, in a sample of whole blood. A hematocrit of 48 means that 48% of the volume of blood consists of RBCs. RBCs transport oxygen to the cells of the body. Therefore, the higher the hematocrit, the more RBCs are present in the blood and the greater the oxygen-carrying potential of the blood. Males usually have higher hematocrit levels than females because males have higher levels of testosterone. In addition to promoting the male sex characteristics, testosterone is responsible for stimulating the release of erythropoietin from the kidneys. Erythropoietin (EPO) is a hormone that stimulates the synthesis of RBCs. Therefore, higher levels of testosterone lead to more EPO secretion and, thus, higher hematocrit levels.

The **erythrocyte sedimentation rate (ESR)** measures the settling of RBCs in a vertical, stationary tube of blood during one hour. In a healthy individual, RBCs do not settle very much in an hour. In some disease conditions, increased production of fibrinogen and immunoglobulins causes the RBCs to clump together, stack up, and form a column (called a *rouleaux formation*). RBCs in a rouleaux formation are heavier and settle faster (that is, they display an increase in the sedimentation rate.)

Hemoglobin (Hb), a protein found in RBCs, is necessary for the transport of oxygen from the lungs to the cells of the body. Four polypeptide chains of amino acids comprise the globin part of the molecule. Each polypeptide chain has a heme unit—a group of atoms that includes an atom of iron to which a molecule of oxygen binds. Each polypeptide chain, if it folds correctly, can bind a molecule of oxygen. Therefore, each hemoglobin molecule can carry four molecules of oxygen. Oxygen combined with hemoglobin forms oxyhemoglobin, which has a bright red color.

All of the cells in the human body, including RBCs, are surrounded by a plasma membrane that contains genetically determined glycoproteins, called antigens. On RBC membranes, there are certain antigens, called **agglutinogens,** that determine a person's blood type. Blood typing is used to identify the **ABO blood groups,** which are determined by the presence or absence of two antigens: **type A** and **type B.** Because these antigens are genetically determined, a person has two copies (alleles) of the gene for these antigens, one copy from each parent.

Cholesterol is a lipid substance that is essential for life—it is an important component of all cell membranes and is the base molecule of steroid hormones, vitamin D, and bile salts. Cholesterol is produced in the human liver and is present in some foods of animal origin, such as milk, meat, and eggs. Because cholesterol is a hydrophobic lipid, it needs to be wrapped in protein packages, called **lipoproteins,** to travel in the blood (which is mostly water) from the liver and digestive organs to the cells of the body.

A C T I V I T Y 1

Hematocrit Determination

OBJECTIVES

1. To understand the terms *hematocrit, red blood cells, hemoglobin, buffy coat, anemia,* and *polycythemia.*

2. To understand how the hematocrit (packed red blood cell volume) is determined.

3. To understand the implications of elevated or decreased hematocrit.

4. To understand the importance of proper disposal of laboratory material that comes in contact with blood.

Introduction

Hematocrit refers to the percentage of **red blood cells (RBCs),** or erythrocytes, in a sample of whole blood. A hematocrit of 48 means that 48% of the volume of blood consists of RBCs. RBCs transport oxygen to the cells of the body. Therefore, the higher the hematocrit, the more RBCs are present in the blood and the higher the oxygen-carrying potential of the blood. Hematocrit values are determined by spinning a microcapillary tube filled with a sample of whole blood in a special microhematocrit centrifuge. This procedure separates the blood cells from the blood plasma. A **buffy**

coat layer of white blood cells (WBCs) appears as a thin, white layer *between* the heavier RBC layer and the lighter, yellow plasma.

The hematocrit is determined after centrifuging by measuring the height of the RBC layer (in millimeters) and dividing that by the height of the total blood sample (in millimeters). This calculation gives the percentage of the total blood volume consisting of RBCs. The average hematocrit for males is 42–52%, and the average hematocrit for females is 37–47%. A lower-than-normal hematocrit indicates **anemia,** and a higher-than-normal hematocrit indicates **polycythemia.**

Anemia is a condition in which insufficient oxygen is transported to the body's cells. There are many possible causes for anemia, including inadequate numbers of RBCs, a decreased amount of the oxygen-carrying pigment **hemoglobin** in the RBCs, and abnormally shaped hemoglobin. The heme portion of a hemoglobin molecule contains an atom of iron to which a molecule of oxygen can bind. If adequate iron is not available, the body cannot manufacture hemoglobin, resulting in the condition *iron-deficiency anemia. Aplastic anemia* results from the failure of the bone marrow to produce adequate red blood cell numbers. *Sickle cell anemia* is an inherited condition in which the protein portion of hemoglobin molecules folds incorrectly when oxygen levels are low. As a result, oxygen molecules cannot bind to the misshapen hemoglobin, the RBCs develop a sickle shape, and anemia results. Regardless of the underlying cause, anemia causes a reduction in the blood's ability to transport oxygen to the cells of the body.

Polycythemia refers to an increase in RBCs, resulting in a higher-than-normal hematocrit. There are many possible causes of polycythemia, including living at high altitudes, strenuous athletic training, and tumors in the bone marrow. In this activity you will simulate the blood test used to determine hematocrit.

> **EQUIPMENT USED** The following equipment will be depicted on-screen: six heparinized capillary tubes (heparin keeps blood from clotting); blood samples from six individuals: sample 1: a healthy male living in Boston, sample 2: a healthy female living in Boston, sample 3: a healthy male living in Denver, sample 4: a healthy female living in Denver, sample 5: a male with aplastic anemia, sample 6: a female with iron-deficiency anemia; capillary tube sealer—a clay material (shown as an orange-yellow substance) used to seal the capillary tubes on one end so the blood sample can be centrifuged without having the blood spray out of the tube; microhematocrit centrifuge—used to centrifuge the samples (rotates at 14,500 revolutions per minute); metric ruler; biohazardous waste disposal—used to properly dispose of equipment that comes in contact with blood.

Experiment Instructions

Go to the home page in the PhysioEx software and click **Exercise 11: Blood Analysis.** Click **Activity 1: Hematocrit Determination,** and take the online **Pre-lab Quiz** for Activity 1.

After you take the online Pre-lab Quiz, click the **Experiment** tab and begin the experiment. The experiment instructions are reprinted here for your reference. The opening screen for the experiment is shown below.

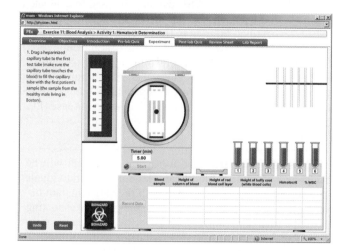

1. Drag a heparinized capillary tube to the first test tube (make sure the capillary tube touches the blood) to fill the capillary tube with the first patient's sample (the sample from the healthy male living in Boston).

2. Drag the capillary tube containing sample 1 to the container of capillary tube sealer to seal one end of the tube.

3. Drag the capillary tube to the microhematocrit centrifuge. The remaining samples will automatically be prepared for centrifugation.

4. Note that the timer is set to 5 minutes. Click **Start** to centrifuge the samples for 5 minutes at 14,500 revolutions per minute. The simulation compresses the 5-minute time period into 5 seconds of real time.

5. Drag capillary tube 1 from the centrifuge to the metric ruler to measure the height of the column of blood and the height of each layer.

6. Click **Record Data** to display your results in the grid (and record your results in Chart 1).

7. Drag capillary tube 1 to the biohazardous waste disposal.

? PREDICT Question 1
Predict how the hematocrits of the patients living in Denver, Colorado (approximately one mile above sea level), will compare with the hematocrit levels of the patients living in Boston, Massachusetts (at sea level).

8. You will now measure the column and layer heights of the remaining samples.

- Drag the next capillary tube from the centrifuge to the metric ruler.

- Click **Record data** to display your results in the grid (and record your results in Chart 1). The tube will automatically be placed in the biohazardous waste disposal.

Repeat this step for each of the remaining samples.

After you complete the experiment, take the online **Post-lab Quiz** for Activity 1.

CHART 1	Hematocrit Determination				
	Total height of column of blood (mm)	Height of red blood cell layer (mm)	Height of buffy coat (mm)	Hematocrit	% WBC
Sample 1 (healthy male living in Boston)					
Sample 2 (healthy female living in Boston)					
Sample 3 (healthy male living in Denver)					
Sample 4 (healthy female living in Denver)					
Sample 5 (male with aplastic anemia)					
Sample 6 (female with iron-deficiency anemia)					

Activity Questions

1. How do you calculate the hematocrit after you centrifuge the total blood sample? What does the result of this calculation indicate?

_____

_____

2. What is the significance of the "buffy coat" after you centrifuge the total blood sample?

_____

_____

3. As noted in the Exercise Overview, the average hematocrit for males is 42–52%, the average hematocrit for females is 37–47%, and erythropoietin is a hormone that is responsible for the synthesis of RBCs. Given this information, explain how a female could have a consistent hematocrit of 48, large, well-defined skeletal muscles, and an abnormally deep voice.

_____

_____

_____ ▬

ACTIVITY 2

Erythrocyte Sedimentation Rate

OBJECTIVES

1. To understand *erythrocyte sedimentation rate (ESR), red blood cells (RBCs),* and *rouleaux formation.*

2. To learn how to perform an erythrocyte sedimentation rate blood test.

3. To understand the results (and their implications) from an erythrocyte sedimentation rate blood test.

4. To understand the importance of proper disposal of laboratory material that comes in contact with blood.

Introduction

The **erythrocyte sedimentation rate (ESR)** measures the settling of **red blood cells (RBCs)** in a vertical, stationary tube of whole blood during one hour. In a healthy individual, red blood cells do not settle very much in an hour. In some disease conditions, increased production of fibrinogen and immunoglobulins cause the RBCs to clump together, stack up, and form a dark red column (called a **rouleaux formation**). RBCs in a rouleaux formation are heavier and settle faster (that is, they exhibit an increase in the settling rate).

The ESR is neither very specific nor diagnostic, but it can be used to follow the progression of certain diseases, including sickle cell anemia, some cancers, and inflammatory diseases, such as rheumatoid arthritis. When the disease worsens, the ESR increases. When the disease improves, the ESR decreases.

The ESR can be elevated in iron-deficiency anemia, and menstruating females sometimes develop anemia and show an increase in ESR. The ESR can also be used to evaluate a patient with chest pains because the ESR is elevated in established myocardial infarction (heart attack) but normal in angina pectoris (chest pain without myocardial infarction). Similarly, it can be useful in screening a female patient with severe abdominal pains because the ESR is not elevated within the first 24 hours of acute appendicitis but is elevated in the early stage of acute pelvic inflammatory disease (PID) or ruptured ectopic pregnancy.

> **EQUIPMENT USED** The following equipment will be depicted on-screen: blood samples from six individuals (each sample has been treated with the anticoagulant heparin): sample 1: healthy individual, sample 2: menstruating female, sample 3: individual with sickle cell anemia, sample 4: individual with iron-deficiency anemia, sample 5: individual suffering a myocardial infarction, sample 6: individual with angina pectoris; sodium citrate—used to bind with calcium and prevent the blood samples from clotting so they can be easily poured into the narrow sedimentation rate tubes; test tubes—used as reaction vessels for the tests; sedimentation tubes (contained in cabinet); magnifying chamber—used to help read the millimeter markings on the sedimentation tubes; biohazardous waste disposal—used to properly dispose of equipment that comes in contact with blood.

Experiment Instructions

Go to the home page in the PhysioEx software and click **Exercise 11: Blood Analysis.** Click **Activity 2: Erythrocyte Sedimentation Rate,** and take the online **Pre-lab Quiz** for Activity 2.

After you take the online Pre-lab Quiz, click the **Experiment** tab and begin the experiment. The experiment instructions are reprinted here for your reference. The opening screen for the experiment is shown below.

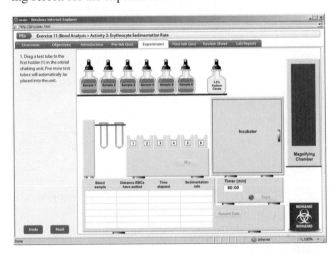

1. Drag a test tube to the first holder (1) in the orbital shaking unit. Five more test tubes will automatically be placed into the unit.

2. Drag the dropper cap of the sample 1 bottle (the sample from the healthy individual) to the first test tube (1) in the orbital shaking unit to dispense one milliliter of blood into the tube. The remaining five samples will be automatically dispensed.

3. Drag the dropper cap of the 3.8% sodium citrate bottle to the first test tube to dispense 0.5 milliliters of sodium citrate into each of the tubes.

4. Click **Mix** to mix the samples.

5. Drag the first test tube to the first sedimentation tube in the incubator to pour the contents of the test tube into the sedimentation tube.

6. Drag the now empty test tube to the biohazardous waste disposal. The contents of the remaining test tubes will automatically be poured into the sedimentation tubes, and the empty tubes will automatically be placed in the biohazardous waste disposal.

7. Note that the timer is set to 60 minutes. Click **Start** to incubate the sedimentation tubes for 60 minutes. The simulation compresses the 60-minute time period into 6 seconds of real time.

8. Drag the first sedimentation tube to the magnifying chamber to examine the tube. The tube is marked in millimeters (the distance between two marks is 5 mm).

9. Click **Record Data** to display your results in the grid (and record your results in Chart 2).

10. Drag the sedimentation tube to the biohazardous waste disposal.

? PREDICT Question 1
How will the sedimentation rate for sample 6 (unhealthy individual) compare with the sedimentation rate for sample 1 (healthy individual)?

11. You will now measure the sedimentation rate for the remaining samples.

- Drag the next sedimentation tube to the magnifying chamber to examine the tube.
- Click **Record Data** to display your results in the grid (and record your results in Chart 2). The tube will automatically be placed in the biohazardous waste disposal.

Repeat this step for each of the remaining samples.

After you complete the experiment, take the online **Post-lab Quiz** for Activity 2.

Activity Questions

1. Why is ESR useful, even though it is neither specific nor sensitive?

_____

_____

2. Describe the physical process underlying an accelerated erythrocyte sedimentation rate.

_____

_____

ACTIVITY 3

Hemoglobin Determination

OBJECTIVES

1. To understand the terms *hemoglobin (Hb)*, *anemia*, *heme*, *oxyhemoglobin*, and *hemoglobinometer*.
2. To learn how to determine the amount of hemoglobin in a blood sample.
3. To understand the results and their implications when examining the amounts of hemoglobin present in a blood sample.
4. To understand the importance of proper disposal of laboratory material that comes in contact with blood.

CHART 2	Erythrocyte Sedimentation Rate		
Blood sample	Distance RBCs have settled (min)	Elapsed time	Sedimentation rate
Sample 1 (healthy individual)			
Sample 2 (menstruating female)			
Sample 3 (individual with sickle cell anemia)			
Sample 4 (individual with iron-deficiency anemia)			
Sample 5 (individual suffering a myocardial infarction)			
Sample 6 (individual with angina pectoris)			

Introduction

Hemoglobin (Hb), a protein found in red blood cells, is necessary for the transport of oxygen from the lungs to the cells of the body. Four polypeptide chains of amino acids comprise the globin part of the molecule. Each polypeptide chain has a **heme** unit—a group of atoms that includes an atom of iron to which a molecule of oxygen binds. Each polypeptide chain, if it folds correctly, can bind a molecule of oxygen. Therefore, each hemoglobin molecule can carry four molecules of oxygen. Oxygen combined with hemoglobin forms **oxyhemoglobin,** which has a bright-red color. Anemia results when insufficient oxygen is carried in the blood.

A quantitative hemoglobin measurement is used to determine the classification and possible causes of anemia and also gives useful information on some other disease conditions. For example, a person can have anemia with a normal red blood cell count if there is inadequate hemoglobin in the red blood cells. Normal blood contains an average of 12–18 grams of hemoglobin per 100 milliliters of blood. A healthy male has 13.5–18 g/100 ml and a healthy female has 12–16 g/100 ml. Hemoglobin levels increase in patients with polycythemia, congestive heart failure, and chronic obstructive pulmonary disease (COPD). Hemoglobin levels also increase when dwelling at high altitudes. Hemoglobin levels decrease in patients with anemia, hyperthyroidism, cirrhosis of the liver, renal disease, systemic lupus erythematosus, and severe hemorrhage.

The hemoglobin level of a blood sample is determined by stirring the blood with a wooden stick to rupture, or lyse, the red blood cells. The color intensity of the hemolyzed blood reflects the amount of hemoglobin present. A **hemoglobinometer** transmits green light through the hemolyzed blood sample and then compares the amount of light that passes through the sample to standard color intensities to determine the hemoglobin content of the sample.

> **EQUIPMENT USED** The following equipment will be depicted on-screen: blood samples from five individuals: sample 1: healthy male, sample 2: healthy female, sample 3: female with iron-deficiency anemia, sample 4: male with polycythemia, sample 5: female Olympic athlete; hemolysis sticks—used to stir the blood samples to lyse the red blood cells, thereby releasing their hemoglobin; blood chamber dispenser—used to dispense a blood chamber slide with a depression for the blood sample; hemoglobinometer—used to analyze the hemoglobin level in each sample; biohazardous waste disposal—used to properly dispose of equipment that comes in contact with blood.

Experiment Instructions

Go to the home page in the PhysioEx software and click **Exercise 11: Blood Analysis.** Click **Activity 3: Hemoglobin Determination,** and take the online **Pre-lab Quiz** for Activity 3.

After you take the online Pre-lab Quiz, click the **Experiment** tab and begin the experiment. The experiment instructions are reprinted here for your reference. The opening screen for the experiment is shown below.

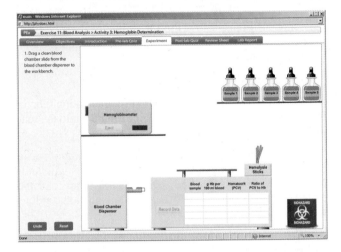

1. Drag a clean blood chamber slide from the blood chamber dispenser to the workbench.

2. Drag the bottle cap from the sample 1 bottle (the sample from the healthy male) to the depression in the blood chamber slide to dispense a drop of blood into the depression.

3. Drag a hemolysis stick to the drop of blood in the chamber to stir the blood sample for 45 seconds, lysing the red blood cells and releasing their hemoglobin.

4. Drag the hemolysis stick to the biohazardous waste disposal.

5. Drag the blood chamber slide to the dark rectangular slot on the hemoglobinometer to analyze the sample. After you insert the blood chamber slide into the hemoglobinometer, you will see a blowup of the inside of the hemoglobinometer.

6. The left half of the circular field shows the intensity of green light transmitted by blood sample 1. The right half of the circular field shows the intensity of green light for known levels of hemoglobin present in blood. Drag the lever on the right side of the hemoglobinometer down until the shade of green in the right half of the field matches the shade of green in the left half of the field and then click **Record Data** to display your results in the grid (and record your results in Chart 3).

7. Click **Eject** to remove the blood chamber slide from the hemoglobinometer.

8. Drag the blood chamber slide from the hemoglobinometer to the biohazardous waste disposal.

> **? PREDICT Question 1**
> How will the hemoglobin levels for the female Olympic athlete (sample 5) compare with the hemoglobin levels for the healthy female (sample 2)?

CHART 3	Hemoglobin Determination		
Blood sample	Hb in grams per 100 ml of blood	Hematocrit (PCV)	Ratio of PCV to Hb
Sample 1 (healthy male)		48	
Sample 2 (healthy female)		44	
Sample 3 (female with iron-deficiency anemia)		40	
Sample 4 (male with polycythemia)		60	
Sample 5 (female Olympic athlete)		60	

9. You will now measure the hemoglobin levels for each of the remaining samples.

- Drag a blood chamber slide to the workbench.

- Drag the bottle cap from the next sample bottle to the depression in the slide.

- Drag a hemolysis stick to the drop of blood in the chamber (after stirring the sample, the hemolysis stick will automatically be placed in the biohazardous waste disposal).

- Drag the blood chamber slide to the dark rectangular slot on the hemoglobinometer.

- Drag the lever on the right side of the hemoglobinometer down until the shade of green in the right half of the field matches the shade of green in the left half of the field and then click **Record Data** to display your results in the grid (and record your results in Chart 3).

- Click **Eject** to remove the blood chamber slide from the hemoglobinometer (the slide will automatically be placed in the biohazardous waste disposal).

Repeat this step until you analyze all five samples.

After you complete the experiment, take the online **Post-lab Quiz** for Activity 3.

Activity Questions

1. As mentioned in the introduction to this activity, hemoglobin levels increase for people living at high altitudes. Given that the atmospheric pressure of oxygen significantly declines as you ascend to higher elevations, why do you think hemoglobin levels would increase for those living at high altitudes?

2. Just by looking at the color of a freshly drawn blood sample, how could you distinguish between blood that is well oxygenated and blood that is poorly oxygenated?

Blood Typing

OBJECTIVES

1. To understand the terms *antigens, agglutinogens, ABO antigens, Rh antigens*, and *agglutinins*.
2. To learn how to perform a blood-typing assay.
3. To understand the results and their implications when examining agglutination reactions.
4. To understand the importance of proper disposal of laboratory material that comes in contact with blood.

Introduction

All of the cells in the human body, including red blood cells, are surrounded by a plasma membrane that contains genetically determined glycoproteins, called **antigens.** On red blood cell membranes, there are certain antigens, called **agglutinogens,** that determine a person's blood type. If a blood transfusion recipient has antibodies (called **agglutinins**) that react with the antigens present on the transfused cells, the red blood cells will become clumped together, or agglutinated, and then lysed, resulting in a potentially life-threatening blood transfusion reaction. It is therefore important to determine an individual's blood type before performing blood transfusions to avoid mixing incompatible blood. Although many different antigens are present on red blood cell membranes, the **ABO** and **Rh antigens** cause the most vigorous and potentially fatal transfusion reactions.

The ABO blood groups are determined by the presence or absence of two antigens: type A and type B. Because these antigens are genetically determined, a person has two copies (alleles) of the gene for these proteins, one copy from each parent. The presence of these antigens is due to a dominant allele, and their absence is due to a recessive allele.

- A person with type A blood can have two alleles for the type A antigen or one allele for the type A antigen and one allele for the absence of either the type A or type B antigen.

- A person with type B blood can have two alleles for the type B antigen or one allele for the type B antigen and one allele for the absence of either the type A or type B antigen.

- A person with type AB blood has one allele for the type A antigen and one allele for the type B antigen.

- A person with type O blood has two recessive alleles and has neither the type A nor type B antigen.

TABLE 11.1	ABO Blood Types	
Blood type	Antigens on RBCs	Antibodies present in plasma
A	A	anti-B
B	B	anti-A
AB	A and B	none
O	none	anti-A and anti-B

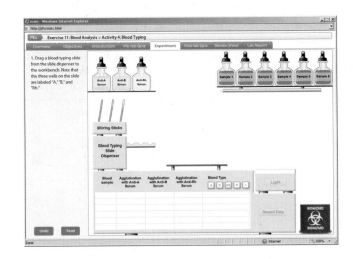

Antibodies against the A and B antigens are found pre-formed in the blood plasma. A person has antibodies only for the antigens not on his or her red blood cells, so a person with type A blood will have anti-B antibodies. View Table 11.1 for a summary of the antigens on red blood cells and the antibodies in the plasma for each blood type.

The Rh factor is another genetically determined protein that can be present on red blood cell membranes. Approximately 85% of the population is Rh positive (Rh$^+$), and their red blood cells have this protein on their surface. Antibodies against the Rh factor are not found preformed in the plasma. They are produced by an Rh negative (Rh$^-$) individual only after exposure to blood cells from someone who is Rh$^+$. Such exposure can occur during pregnancy when Rh$^+$ blood cells from the baby cross the placenta and expose the mother to the antigen.

To determine an individual's blood type, drops of an individual's blood sample are mixed separately with antiserum containing antibodies to either type A antigens, type B antigens, or Rh antigens. An agglutination reaction (showing clumping) indicates the presence of the agglutinogen.

EQUIPMENT USED The following equipment will be depicted on-screen: blood samples from six individuals with different blood types; anti-A serum (blue bottle), anti-B serum (yellow bottle), and anti-Rh serum (white bottle), containing antibodies to the A antigen, B antigen, and Rh antigen, respectively; blood-typing slide dispenser; color-coded stirring sticks—used to mix the blood sample and the serum (blue: used with anti-A serum, yellow: used with the anti-B serum, white: used with the anti-Rh serum); light box—used to view the blood type samples; biohazardous waste disposal—used to properly dispose of equipment that comes in contact with blood.

Experiment Instructions

Go to the home page in the PhysioEx software and click **Exercise 11, Blood Analysis.** Click **Activity 4, Blood Typing,** and take the online **Pre-lab Quiz** for Activity 4.

After you take the online Pre-lab Quiz, click the **Experiment** tab and begin the experiment. The experiment instructions are reprinted here for your reference. The opening screen for the experiment is shown above.

1. Drag a blood-typing slide from the slide dispenser to the workbench. Note that the three wells on the slide are labeled "A," "B," and "Rh."

2. Drag the dropper cap of the sample 1 bottle to well A on the blood-typing slide to dispense a drop of blood into each well.

3. Drag the dropper cap of the anti-A serum bottle to well A on the blood-typing slide to dispense a drop of anti-A serum into the well.

4. Drag the dropper cap of the anti-B serum bottle to well B on the blood-typing slide to dispense a drop of anti-B serum into the well.

5. Drag the dropper cap of the anti-Rh serum bottle to well Rh on the blood-typing slide to dispense a drop of anti-Rh serum into the well.

6. Drag a blue-tipped stirring stick to well A to mix the blood and anti-A serum.

7. Drag the stirring stick to the biohazardous waste disposal.

8. Drag a yellow-tipped stirring stick to well B to mix the blood and anti-B serum.

9. Drag the stirring stick to the biohazardous waste disposal.

10. Drag a white-tipped stirring stick to well Rh to mix the blood and anti-Rh serum.

11. Drag the stirring stick to the biohazardous waste disposal.

12. Drag the blood-typing slide to the light box and then click **Light** to analyze the slide.

13. Under each of the wells, click **Positive** if agglutination occurred (the sample shows clumping) or click **Negative** if agglutination did not occur (the sample looks smooth).

CHART 4	Blood Typing Results			
Blood sample	Agglutination with anti-A serum	Agglutination with anti-B serum	Agglutination with anti-Rh serum	Blood type
1				
2				
3				
4				
5				
6				

14. Click **Record Data** to display your results in the grid (and record your results in Chart 4).

15. Drag the blood-typing slide to the biohazardous waste disposal.

? PREDICT Question 1
If the patient's blood type is AB⁻, what would be the appearance of the A, B, and Rh samples?

16. You will now analyze the remaining samples.

• Drag a blood-typing slide from the slide dispenser to the workbench. The next sample will be added to each well on the slide, the appropriate antiserum will be added to each well, the sample and antisera will be mixed, and the slide will be placed in the light box.

• Under each of the wells, click **Positive** if agglutination occurred (the sample shows clumping) or click **Negative** if agglutination did not occur (the sample looks smooth).

• Click **Record Data** to display your results in the grid (and record your results in Chart 4).

Repeat this step until you analyze all six samples.

17. You will now indicate the blood type for each sample and indicate whether the sample is Rh positive or Rh negative.

• Click the row for the sample in the grid (and record your results in Chart 4).

• Click A, B, AB, or O above the blood type column to indicate the blood type.

• Click the − button or the + button above the blood type column to indicate whether the sample is Rh negative or Rh positive.

Repeat this step for all six samples. Record your results in Chart 4.

After you complete the experiment, take the online **Post-lab Quiz** for Activity 4.

Activity Questions

1. Antibodies against the A and B antigens are found in the plasma, and a person has antibodies only for the antigens that are not present on their red blood cells. Using this information, list the antigens found on red blood cells and the antibodies in the plasma for blood types 1) AB−, 2) O+, 3) B−, and 4) A+.

_____

_____

2. If an individual receives a bone marrow transplant from someone with a different ABO blood type, what happens to the recipient's ABO blood type?

_____

_____

Blood Cholesterol

OBJECTIVES

1. To understand the terms *cholesterol, lipoproteins, low-density lipoprotein (LDL), hypocholesterolemia, hypercholesterolemia,* and *atherosclerosis.*

2. To learn how to test for total blood cholesterol using a colorimetric assay.

3. To understand the results and their implications when examining total blood cholesterol.

4. To understand the importance of proper disposal of laboratory material that comes in contact with blood.

Introduction

Cholesterol is a lipid substance that is essential for life—it is an important component of all cell membranes and is the base molecule of steroid hormones, vitamin D, and bile salts. Cholesterol is produced in the human liver and is present in some foods of animal origin, such as milk, meat, and eggs. Because cholesterol is a water-insoluble lipid, it needs to be wrapped in protein packages, called **lipoproteins,** to travel in the blood (which is mostly water) from the liver and digestive organs to the cells of the body.

One type of lipoprotein package, called **low-density lipoprotein (LDL),** has been identified as a potential source of damage to the interior of arteries. LDLs can contribute to **atherosclerosis,** the buildup of plaque, in these blood vessels.

A total blood cholesterol determination does not measure the level of LDLs, but it does provide valuable information about the total amount of cholesterol in the blood.

Less than 200 milligrams of total cholesterol per deciliter of blood is considered desirable. Between 200 and 239 mg/dl is considered borderline high cholesterol. Over 240 mg/dl is considered high blood cholesterol (**hypercholesterolemia**) and is associated with an increased risk of cardiovascular disease. Abnormally low blood cholesterol levels (total cholesterol lower than 100 mg/dl) can also suggest a problem. Low levels may indicate hyperthyroidism (overactive thyroid gland), liver disease, inadequate absorption of nutrients from the intestine, or malnutrition. Other reports link **hypocholesterolemia** (low blood cholesterol) to depression, anxiety, and mood disturbances, which are thought to be controlled by the level of available serotonin, a neurotransmitter. There is evidence of a relationship between low levels of blood cholesterol and low levels of serotonin in the brain.

In this test for total blood cholesterol, a sample of blood is mixed with enzymes that produce a colored reaction with cholesterol. The intensity of the color indicates the amount of cholesterol present. The cholesterol tester compares the color of the sample to the colors of known levels of cholesterol (standard values).

EQUIPMENT USED The following equipment will be depicted on-screen: lancets—sharp, needlelike instruments used to prick the finger to obtain a drop of blood; four patients (represented by an extended finger); alcohol wipes—used to cleanse the patient's fingertip before it is punctured with the lancet; color wheel—divided into shades of green that correspond to total cholesterol levels; cholesterol strips—contain chemicals that convert, by a series of reactions, the cholesterol in the blood sample into a green-colored solution; biohazardous waste disposal—used to properly dispose of equipment that comes in contact with blood.

Experiment Instructions

Go to the home page in the PhysioEx software and click **Exercise 11: Blood Analysis.** Click **Activity 5: Blood Cholesterol,** and take the online **Pre-lab Quiz** for Activity 5.

After you take the online Pre-lab Quiz, click the **Experiment** tab and begin the experiment. The experiment instructions are reprinted here for your reference. The opening screen for the experiment is shown below.

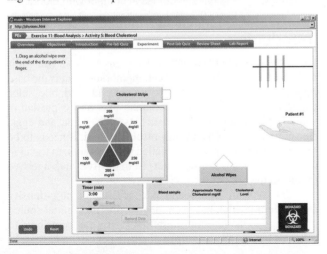

1. Drag an alcohol wipe over the end of the first patient's finger.

2. Drag the alcohol wipe to the biohazardous waste disposal.

3. Drag a lancet to the tip of the patient's finger to prick the finger and obtain a drop of blood.

4. Drag the lancet to the biohazardous waste disposal.

5. Drag a cholesterol strip to the finger to transfer a drop of blood from the patient's finger to the strip.

6. Drag the cholesterol strip to the rectangular box to the right of the color wheel.

7. Click **Start** to start the timer. It takes three minutes for the chemicals in the cholesterol strip to react with the blood. The simulation compresses the 3-minute time period into 3 seconds of real time.

8. Click the color on the color wheel that most closely matches the color on the cholesterol strip.

9. Click **Record Data** to display your results in the grid (and record your results in Chart 5).

CHART 5	Total Cholesterol Determination	
Blood sample	Approximate total cholesterol (mg/dl)	Cholesterol level
1		
2		
3		
4		

10. Drag the cholesterol test strip to the biohazardous waste disposal.

 PREDICT Question 1
Patient 4 prefers to cook all his meat in lard or bacon grease. Knowing this dietary preference, you anticipate his total cholesterol level to be:

_____

11. You will now test the total cholesterol levels for the remaining patients.

- Drag an alcohol wipe over the end of the patient's finger. The alcohol wipe will automatically be placed in the biohazardous waste disposal.

- Drag a lancet to the tip of the patient's finger to prick the finger and obtain a drop of blood. The lancet will automatically be placed in the biohazardous waste disposal.

- Drag a cholesterol strip to the finger to transfer a drop of blood from the patient's finger to the strip.

- Drag the cholesterol strip to the rectangular box to the right of the color wheel. The timer will automatically run for three minutes to allow the chemicals in the cholesterol strip to react with the blood.

- Click the color on the color wheel that most closely matches the color on the cholesterol strip.

- Click **Record Data** to display your results in the grid (and record your results in Chart 5). The cholesterol strip will automatically be placed in the biohazardous waste disposal.

    Repeat this step until you determine the total cholesterol levels for all four patients.

After you complete the experiment, take the online **Post-lab Quiz** for Activity 5.

Activity Questions

1. Why do cholesterol plaques occur in arteries and not veins?

_____

_____

2. Phytosterols can alter absorption of certain molecules by the intestinal tract. Why would they be a beneficial dietary supplement for people with high LDL levels?

_____

_____

Blood Analysis

NAME_____

LAB TIME/DATE_____

ACTIVITY 1 Hematocrit Determination

1. List the hematocrits for the healthy male (sample 1) and female (sample 2) living in Boston (at sea level) and indicate whether they are normal or whether they indicate anemia or polycythemia.

   _____

   _____

2. Describe the difference between the hematocrits for the male and female living in Boston. Why does this difference between the sexes exist?

   _____

   _____

3. List the hematocrits for the healthy male and female living in Denver (approximately one mile above sea level) and indicate whether they are normal or whether they indicate anemia or polycythemia.

   _____

   _____

4. How did the hematocrit levels of the Denver residents differ from those of the Boston residents? Why? How well did the results compare with your prediction?

   _____

   _____

5. Describe how the kidneys respond to a chronic decrease in oxygen and what effect this has on hematocrit levels.

   _____

   _____

6. List the hematocrit for the male with aplastic anemia (sample 5) and indicate whether it is normal or abnormal. Explain your response.

   _____

   _____

7. List the hematocrit for the female with iron-deficiency anemia (sample 6) and indicate whether it is normal or abnormal. Explain your response.

_____

_____

A C T I V I T Y 2 Erythrocyte Sedimentation Rate

1. Describe the effect that sickle cell anemia has on the sedimentation rate (sample 3). Why do you think that it has this effect?

_____

_____

2. How did the sedimentation rate for the menstruating female (sample 2) compare with the sedimentation rate for the healthy individual (sample 1)? Why do you think this occurs?

_____

_____

3. How did the sedimentation rate for the individual with angina pectoris (sample 6) compare with the sedimentation rate for the healthy individual (sample 1)? Why? How well did the results compare with your prediction?

_____

_____

4. What effect does iron-deficiency anemia (sample 4) have on the sedimentation rate?

_____

_____

5. Compare the sedimentation rate for the individual suffering a myocardial infarction (sample 5) with the sedimentation rate for the individual with angina pectoris (sample 6). Explain how you might use this data to monitor heart conditions.

_____

_____

A C T I V I T Y 3 Hemoglobin Determination

1. Is the male with polycythemia (sample 4) deficient in hemoglobin? Why?

_____

_____

2. How did the hemoglobin levels for the female Olympic athlete (sample 5) compare with the hemoglobin levels for the healthy female (sample 2)? Is either person _deficient_ in hemoglobin? How well did the results compare with your prediction?

_____

_____

3. List conditions in which hemoglobin levels would be expected to decrease. Provide reasons for the change when possible.

_____

_____

4. List conditions in which hemoglobin levels would be expected to increase. Provide reasons for the change when possible.

_____

_____

5. Describe the ratio of hematocrit to hemoglobin for the healthy male (sample 1) and female (sample 2). (A normal ratio of hematocrit to grams of hemoglobin is approximately 3:1.) Discuss any differences between the two individuals.

_____

_____

6. Describe the ratio of hematocrit to hemoglobin for the female with iron-deficiency anemia (sample 3) and the female Olympic athlete (sample 5). (A normal ratio of hematocrit to grams of hemoglobin is approximately 3:1.) Discuss any differences between the two individuals.

_____

_____

ACTIVITY 4   Blood Typing

1. How did the appearance of the A, B, and Rh samples for the patient with AB⁻ blood type compare with your prediction?

_____

_____

2. Which blood sample contained the rarest blood type?

_____

3. Which blood sample contained the universal donor?

_____

4. Which blood sample contained the universal recipient?

_____

5. Which blood sample did not agglutinate with any of the antibodies tested? Why?

_____

_____

6. What antibodies would be found in the plasma of blood sample 1?

_____

7. When transfusing an individual with blood that is compatible but not the same type, it is important to separate packed cells from the plasma and administer only the packed cells. Why do you think this is done? (Hint: Think about what is *in plasma* versus what is *on RBCs*.)

8. List the blood samples in this activity that represent people who could donate blood to a person with type B^+ blood.

A C T I V I T Y   5   Blood Cholesterol

1. Which patient(s) had desirable cholesterol level(s)?

2. Which patient(s) had elevated cholesterol level(s)?

3. Describe the risks for the patient(s) you identified in question 2.

4. Was the cholesterol level for patient 4 low, desirable, or high? How well did the results compare with your prediction? What advice about diet and exercise would you give to this patient? Why?

5. Describe some reasons why a patient might have abnormally low blood cholesterol.

Serological Testing

Exercise Overview

Immunology, the study of the immune system, focuses on chemical interactions that are difficult to observe. A number of chemical techniques have been developed to visually represent antibodies and antigens in the **serum,** the fluid portion of the blood with the clotting factors removed. The study and use of these techniques is referred to as **serology.** These techniques are performed in vitro, outside of the body, and are primarily used as diagnostic tools to detect disease. Other applications include pregnancy testing and drug testing. These immunological techniques depend upon the principle that an antibody binds only to specific, corresponding antigens. The tests are relatively expensive to perform, so these activities will allow you to perform them without the sometimes cost-prohibitive supplies.

Antigens and Antibodies

The word **antigen** is derived from two words: *anti*body and *gen*erator. Antigens do not produce antibodies, but early scientists noted that when antigens were present, antibodies appeared. Plasma cells actually produce antibodies.

Antigens include proteins, polysaccharides, and various small molecules that stimulate antibody production. Antigens are often molecules that are described as **nonself,** or foreign to the body. There are also self-antigens that act as identifier tags, such as the proteins found on the surface of red blood cells.

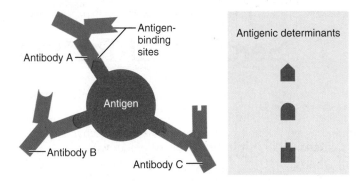

FIGURE 12.1 **Antigen-antibody interaction with antigenic determinants.**

Most often, antigens are a portion of an infectious agent, such as a bacterium or a virus, and the body produces antibodies in response to the presence of the infectious agent.

Antigens are often large and have multiple antigenic sites—locations that can bind to antibodies. We refer to these sites as **antigenic determinants,** or **epitopes.** The antibody has a corresponding antigen-binding site that has a "lock-and-key" recognition for the antigenic determinant on the antigen (view Figure 12.1). All of the simulated tests presented in this exercise take advantage of antigen-antibody specificity. These tests include direct fluorescent antibody technique, Ouchterlony technique, ELISA (enzyme-linked immunosorbent assay), and Western blotting technique.

Nonspecific Binding

The lock-and-key recognition that antigen and antibody have for each other is much like the specificity that an enzyme and its substrate have for one another. However, with antigen and antibody, **nonspecific binding** sometimes occurs. For this reason you will perform a number of washing steps in this exercise to remove any nonspecific binding.

Positive and Negative Controls

You will also use **positive** and **negative controls** to ensure that the test is working accurately. Positive controls include a substance that is known to react positively, thus giving you a standard against which to base your results. Negative controls include substances that should not react. A positive result with a negative control is a "false positive," which would invalidate all other results. Likewise, a negative result with a positive control is a "false negative," which would also invalidate your results.

A C T I V I T Y   1

Using Direct Fluorescent Antibody Technique to Test for Chlamydia

OBJECTIVES

1. To understand how fluorescent antibodies can be used diagnostically to detect the presence of a specific antigen.
2. To observe how to test for the sexually transmitted disease chlamydia.
3. To distinguish between antigens and antibodies.

4. To understand the terms *epitope* and *antigenic determinant.*
5. To observe nonspecific binding that can result between antigen and antibody.

Introduction

The direct fluorescent antibody technique uses antibodies to directly detect the presence of antigen. A fluorescent dye molecule attached to these antibodies acts as a visual signal for a positive result. This technique is typically used to test for antigens from infectious agents, such as bacteria or viruses. In this activity you will test for the presence of *Chlamydia trachomatis* (a bacterium that invades the cells of its host) using fluorescently labeled antibodies to detect the presence of the antigen and, therefore, the bacterium. *Chlamydia trachomatis* is an important infectious agent because it causes the sexually transmitted disease **chlamydia.** Left untreated, chlamydia can lead to sterility in men and women.

Chlamydia trachomatis is an obligate, intracellular bacterium, which means that it can only survive inside a host cell. The life cycle of the bacterium has two cellular types. The infectious cell type is the small, dense **elementary body,** which is capable of attaching to the host cell. The **reticulate body** is a larger, less-dense cell, which divides actively once inside the host cell. The reticulate body is also referred to as the vegetative form. The life cycle of *Chlamydia* begins when the elementary body enters the host cell and continues as the elementary body changes inside the host cell into a reticulate body. The reticulate body divides into more reticulate bodies and converts back to the elementary body form for release to infect other cells.

In this activity you will test three patient samples and two control samples for the *Chlamydia* infection. An epithelial scraping from the male urethra or from the cervix of the uterus is performed to collect squamous cells from the surface. The elementary bodies are measured by reacting antigen-specific antibodies to infected cells. The fluorescent dye attached to the antigen-specific antibodies makes the complex detectable. The sample is viewed with a fluorescent microscope. The presence of ten or more elementary bodies in a field of view with a diameter of 5 millimeters is considered a positive result. The elementary bodies will be stained green inside red host cells.

EQUIPMENT USED The following equipment will be depicted on-screen: five samples: patient A, patient B, patient C, a positive control, and a negative control; incubator; fluorescent microscope; 95% ethyl alcohol—used for fixing the sample to the microscope slide; chlamydia fluorescent antibody (Chlamydia FA)—antibodies specific for the *Chlamydia* antigen with a fluorescent dye attached; fluorescent antibody mounting media (FA mounting)—used to mount the prepared sample to the slide when ready for viewing under the microscope; phosphate buffered saline (PBS)—used to wash off excess antibodies and prevent nonspecific binding of the antigen and antibody; fluorescent antibody buffer (FA buffer)—used to remove excess ethyl alcohol; petri dishes—used for incubation of the slides to keep them moist; microscope slides—an incubation vessel where the antigen and antibody react; cotton-tipped applicators—used for application and mixing of the antibodies with the samples; filter paper—used to keep the samples moist in the petri dishes; biohazardous waste disposal.

Experiment Instructions

Go to the home page in the PhysioEx software and click **Exercise 12: Serological Testing.** Click **Activity 1: Using Direct Fluorescent Antibody Technique to Test for Chlamydia,** and take the online **Pre-lab Quiz** for Activity 1.

After you take the online Pre-lab Quiz, click the **Experiment** tab and begin the experiment. The experiment instructions are reprinted here for your reference. The opening screen for the experiment is shown below.

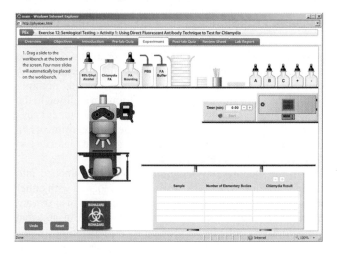

1. Drag a slide to the workbench at the bottom of the screen. Four more slides will automatically be placed on the workbench.

2. The patient samples have been suspended in a small amount of buffer and placed in dropper bottles for ease of dispensing. Drag the dropper cap of the patient A sample bottle to the first slide on the workbench to dispense a drop of the sample onto the slide. A drop from each sample will be placed on a separate slide.

3. Drag the dropper cap of the 95% ethyl alcohol bottle to the first slide on the workbench to dispense three drops of ethyl alcohol onto each slide.

4. Set the timer to 5 minutes by clicking the + button beside the timer display. Click **Start** to start the timer and allow the ethyl alcohol to fix the sample to the slide and prevent the sample from being washed off in the subsequent washing steps. The simulation compresses the 5-minute time period into 5 seconds of real time.

5. Drag the fluorescent antibody (FA) buffer squirt bottle to the first slide to rinse all five slides and remove excess ethyl alcohol.

6. Drag an applicator stick to the chlamydia fluorescent antibody (FA) bottle to soak its cotton tip with antibodies that are specific for *Chlamydia* and labeled with a fluorescent tag.

7. Drag the applicator stick to the first slide to apply the chlamydia fluorescent antibody. Separate applicator sticks will automatically be soaked in chlamydia fluorescent antibody and applied to each slide. Each applicator will automatically be placed in the biohazardous waste disposal.

8. Drag a petri dish to the workbench. A piece of filter paper will be placed into the petri dish. The filter paper has been moistened with fluorescent antibody buffer to keep the

samples from drying out during incubation. Four more petri dishes (and filter paper moistened with fluorescent antibody buffer) will automatically be placed on the workbench.

9. Drag the first slide into the first petri dish. The remaining four slides will automatically be placed into the remaining petri dishes and all five petri dishes will be loaded into the incubator.

10. Set the timer to 20 minutes by clicking the + button next to the timer display. Click **Start** to incubate the samples at 25°C. During incubation the antibodies will react with the corresponding antigens if they are present in the sample. The petri dishes will automatically be removed from the incubator when the time is complete. The simulation compresses the 20-minute incubation time period into 10 seconds of real time.

11. Drag the phosphate buffered saline (PBS) squirt bottle to the first petri dish to wash off excess antibodies and prevent nonspecific binding of the antigen and antibody. The timer will count down 10 minutes for a thorough washing.

12. Click the first petri dish to open the dish and remove the slide. The slides will automatically be removed from the remaining petri dishes.

13. Drag the first petri dish to the biohazardous waste disposal. The remaining petri dishes will automatically be placed in the biohazardous waste disposal.

14. Drag the dropper cap of the fluorescent antibody (FA) mounting media to the first slide to dispense a drop of mounting media onto each slide to mount the sample to the slide.

15. Drag the first slide (patient A) to the fluorescent microscope. Count the number of elementary bodies you see through the microscope (recall that elementary bodies stain green). Click **Submit** to display your results in the grid (and record your results in Chart 1). After you click **Submit,** the slide will automatically be placed in the biohazardous waste disposal.

CHART 1	Direct Fluorescent Antibody Technique Results	
Sample	Number of elementary bodies	Chlamydia result
Patient A		
Patient B		
Patient C		
Positive control		
Negative control		

16. Repeat step 15 for Patient B.

17. Repeat step 15 for Patient C.

18. Repeat step 15 for the Positive Control.

19. Repeat step 15 for the Negative Control.

20. You will now indicate whether each sample is negative or positive for *Chlamydia*. Click the row for the sample in the grid and then click the − button or the + button above the Chlamydia result column to indicate whether the sample is negative or positive for *Chlamydia*. Repeat this step for all five samples. Record your results in Chart 1.

After you complete the experiment, take the online **Post-lab Quiz** for Activity 1.

Activity Questions

1. With this technique, is the antigen or antibody found on the patient sample? Explain how you know this.

_____

_____

2. Explain the difference between an antigen and an epitope (antigenic determinant).

_____

_____

3. When a sample has a small number of elementary bodies but not enough to be a positive result, there appears to have been some nonspecific binding that was not removed by the washing steps. Which sample displayed this property?

_____

_____

ACTIVITY 2

Comparing Samples with Ouchterlony Double Diffusion

OBJECTIVES

1. To observe the precipitation reaction between antigen and antibody.

2. To distinguish between *epitope* and *antigen*.

3. To understand the specificity that antibodies have for their epitopes.

4. To observe how related proteins might share epitopes in common.

Introduction

The Ouchterlony technique is also known as double diffusion. In this technique antigen and antibody diffuse toward each other in a semisolid medium made up of clear, clarified agar. When the antigen and antibody are in optimal proportions, cross-linking of the antigen and antibody occurs, forming an insoluble precipitate, called a **precipitin line.** These lines can

then be used to visually identify similarities between antigens. If optimum proportions have not been met—for example, if there is excess antigen or excess antibody—then no visible precipitate will form. This technique provides easily visible evidence of the binding between antigen and antibody, and sophisticated equipment is not needed to observe the antigen-antibody reaction.

The Ouchterlony technique is designed to determine whether antigens are identical, related, or unrelated. Antigens have **identity** if they are identical. Identical antigens have all their antigenic determinants, or epitopes, in common. In the case of identity, precipitin lines diffuse into each other to completely fuse and form an arc. Antigens have **partial identity** if they are similar or related. Related antigens have some, but not all, antigenic determinants in common. In the case of partial identity, a spur pointing toward the more similar antigen well forms in addition to the arc. Antigens have **non-identity** if they are unrelated. Unrelated antigens do not have any antigenic determinants in common. In the case of non-identity, the lines intersect to form two spurs that resemble an X.

In the Ouchterlony technique, holes are punched into the agar to form wells. The wells are then loaded with either antigen or antibody, which are allowed to diffuse toward each other. Often, the same antigen is placed in adjacent wells to assess the purity of an antigen preparation. In this case a smooth arc with no spurs should be seen, as the antigens are identical. Multiple antibodies can also be placed in a center well. The antibodies will diffuse out in all directions and react with the antigens that are placed in the surrounding wells.

In this activity you will use human and bovine (from cows) albumin as the antigens, and the antibodies will be made in goats against albumin from either humans or cows. The goals are to identify an unknown antigen and to observe the patterns produced by the various relationships: identity, partial identity, and non-identity.

EQUIPMENT USED The following equipment will be depicted on-screen: goat anti–human albumin (Goat A-H)—an antiserum containing antibodies produced by goats against human albumin; goat anti–bovine albumin (Goat A-B)—an antiserum containing antibodies produced by goats against bovine (cow) albumin; bovine serum albumin (BSA); human serum albumin (HSA); unknown antigen; petri dishes filled with clear agar; well cutter.

Experiment Instructions

Go to the home page in the PhysioEx software and click **Exercise 12: Serological Testing.** Click **Activity 2: Comparing Samples with Ouchterlony Double Diffusion:** and take the online **Pre-lab Quiz** for Activity 2.

After you take the online Pre-lab Quiz, click the **Experiment** tab and begin the experiment. The experiment instructions are reprinted here for your reference. The opening screen for the experiment is shown on the following page.

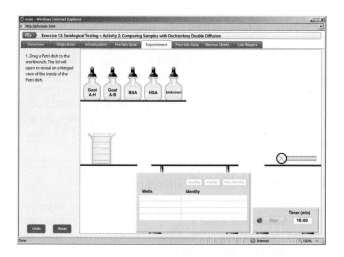

1. Drag a petri dish to the workbench. The lid will open to reveal an enlarged view of the inside of the petri dish.

2. Drag the well cutter to the middle of the enlarged view of the petri dish to punch a hole in the agar in the middle of the petri dish. Drag the well cutter to the upper left, upper right, lower left, and lower right of the petri dish to punch four more holes in the agar. After you punch all five wells into the agar, the wells will be labeled 1–5.

3. Drag the dropper cap of the goat anti–human albumin (Goat A-H) bottle to well 1 to fill it with a sample.

4. Drag the dropper cap of the goat anti–bovine albumin (Goat A-B) bottle to well 1 to fill it with a sample.

5. Drag the dropper cap of the bovine serum albumin (BSA) bottle to well 2 to fill it with a sample.

6. Drag the dropper cap of the bovine serum albumin (BSA) bottle to well 3 to fill it with a sample.

7. Drag the dropper cap of the human serum albumin (HSA) bottle to well 4 to fill it with a sample.

8. Drag the dropper cap of the unknown antigen bottle to well 5 to fill it with a sample.

? PREDICT Question 1
How do you think human serum albumin and bovine serum albumin will compare?

9. Note that the timer is set to 16 hours. Click **Start** to start the timer. The antigen and antibodies will diffuse toward each other and form a precipitate, detected as a precipitin line. The simulation compresses the 16-hour time period into 10 seconds of real time.

10. You will now examine the precipitin lines that formed and indicate the relationship between each pair of antigens. Click the row for the wells containing the antigens in the grid and then click **Identity, Partial,** or **Non-Identity** above the identity column to indicate whether the antigens have identity, partial identity, or non-identity. Repeat this step for the four pairs of antigens. Record your results in Chart 2.

CHART 2	Ouchterlony Double Diffusion Results
Wells	Identity
2 and 5	
2 and 3	
3 and 4	
4 and 5	

After you complete the Experiment, take the online **Post-lab Quiz** for Activity 2.

Activity Questions

1. Which type of identity was present between the samples in this activity? Describe this type of identity.

2. Describe the importance of what you place in the center well.

3. Why do you think it is important for the agar to be clear and clarified?

4. Describe the role that albumin plays in the blood.

ACTIVITY 3

Indirect Enzyme-Linked Immunosorbent Assay (ELISA)

OBJECTIVES

1. To understand how the enzyme-linked immunosorbent assay (ELISA) is used as a diagnostic test.
2. To distinguish between the direct and the indirect ELISA.
3. To describe the basic structure of antibodies.
4. To define *seroconversion.*
5. To understand how the indirect ELISA is used to detect antibodies against HIV.

Introduction

The **enzyme-linked immunosorbent assay (ELISA)** is used to test for the presence of an antigen or antibody. The assay is considered enzyme linked because an enzyme is chemically linked to an antibody in both the direct and indirect versions of the test. Immunosorbent refers to the fact that either antigens or antibodies are being adsorbed (stuck) to plastic. If the test is designed to detect an antigen or antigens, it is a **direct ELISA** because it is directly looking for the foreign substance. An **indirect ELISA** is designed to detect antibodies that the patient has made against the antigen. A positive result with the indirect ELISA requires **seroconversion.** Seroconversion occurs when a patient goes from testing negative for a specific antibody to testing positive for the same antibody.

In the direct ELISA, a 96-well microtiter plate is coated with homologous antibodies made against the antigen of interest. The number of wells makes it easy to test many samples at the same time. The patient serum sample is added to the plate to test for the presence of the antigen that binds to the antibody coating on the plate. ELISA takes advantage of the fact that protein sticks well to plastic. A secondary antibody is added to the plate after the patient serum sample is added. If the antigen is present, a "sandwich" of antibody, antigen, and secondary antibody will form. The secondary antibody is chemically linked to an enzyme. When the substrate is added, the enzyme converts the substrate from a colorless compound to a colored compound. The amount of color produced will be proportional to the amount of antigen binding to the antibodies and thus indicates whether the patient is positive for the antigen. If the antigen is not present, the secondary (enzyme-linked) antibodies will be rinsed away with the washing steps and the substrate will not be converted and will remain colorless. A common use of the direct ELISA is a home pregnancy test, which detects human chorionic gonadotropin (hCG), a hormone present in the urine of pregnant women.

In the indirect ELISA, a 96-well microtiter plate is coated with antigens. The patient serum sample is added to test for the presence of antibodies that bind to the antigens on the plate. The secondary antibody that is added has an enzyme linked to it that binds to the **constant region** of the primary antibody if it is present in the patient sample. The constant region of an antibody has the same sequence of amino acids within a class of antibodies (for example, all IgG antibodies have the same constant region). The **variable region** of an antibody provides the diversity of antibodies and is the site to which the antigen binds. The configuration that forms in the indirect ELISA is antigen, primary antibody, and secondary antibody. Just as in the direct ELISA, the addition of substrate is used to determine whether the sample is positive for the presence of antibody.

In this activity you will use the indirect ELISA to test for the presence of antibodies made against human immunodeficiency virus (HIV). You will use positive and negative controls to verify the results. You will note that an indeterminate result can be obtained if there is not enough color produced to warrant a positive result. The cause of an indeterminate result could be either nonspecific binding or that the individual has been recently infected and has not yet produced enough antibodies for a positive result. In either case, the individual would be retested.

EQUIPMENT USED The following equipment will be depicted on-screen: five samples in the samples cabinet: patient A, patient B, patient C, a positive control, and a negative control; 96-well microtiter plate; multichannel pipettor; 100-μl pipettor; microtiter plate reader; pipettor tip dispenser; washing buffer; HIV antigen solution; developing buffer—secondary antibody conjugated with an enzyme; substrate solution; paper towels—used for blotting; biohazardous waste disposal.

Experiment Instructions

Go to the home page in the PhysioEx software and click **Exercise 12: Serological Testing.** Click **Activity 3: Enzyme-Linked Immunosorbent Assay (ELISA),** and take the online **Pre-lab Quiz** for Activity 3.

After you take the online Pre-lab Quiz, click the **Experiment** tab and begin the experiment. The experiment instructions are reprinted here for your reference. The opening screen for the experiment is shown below.

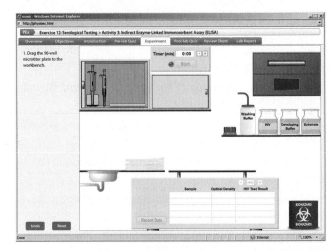

1. Drag the 96-well microtiter plate to the workbench.

2. Drag the multichannel pipettor to the pipette tip dispenser to insert the tips.

3. Drag the multichannel pipettor to the HIV antigens bottle to draw the antigen solution into the tips.

4. Drag the multichannel pipettor directly over the microtiter plate to dispense the liquid into the wells in one row of the plate.

5. Drag the multichannel pipettor to the biohazardous waste disposal for removal and disposal of the tips.

6. Set the timer to 14 hours by clicking the **+** button beside the timer display. This incubation time allows the antigens to stick to the plastic wells of the microtiter plate. Click **Start** to start the timer. The simulation compresses the 14-hour time period into 10 seconds of real time.

7. Drag the washing buffer squeeze bottle to the microtiter plate to remove excess antigens that are not adsorbed (stuck) to the plate.

8. Drag the microtiter plate to the sink to dump the contents of the tray into the sink to remove the washing buffer and excess antigens that are not stuck to the plastic.

9. Drag the microtiter plate to the paper towels. The plate will be pressed to the surface of the paper towels to remove the remaining liquid from the wells. In a typical ELISA, you would perform multiple washing steps to reduce any nonspecific binding. The number of washing steps in this simulation has been reduced for simplicity.

10. Drag the 100-µl pipettor to the tip dispenser to place a tip onto the pipettor.

11. Drag the 100-µl pipettor to the test tube containing the positive control sample (+) to draw the sample into the tip.

12. Drag the 100-µl pipettor to the microtiter plate to dispense the sample into the wells of the plate. The tip will automatically be removed and disposed of in the biohazardous waste disposal. Each of the remaining samples will automatically be dispensed into plate.

13. Set the timer to 1 hour by clicking the + button beside the timer display. This incubation time allows the antigens stuck to the plastic to bind to the antibodies present in the sample. Click **Start** to start the timer. The simulation compresses the 1-hour time period into 10 seconds of real time.

14. Drag the washing buffer squeeze bottle to the microtiter plate to wash off excess antibodies and prevent nonspecific binding of the antigen and antibody.

15. Drag the microtiter plate to the sink to dump washing buffer and unbound antibodies into the sink.

16. Drag the microtiter plate to the paper towels. The plate will be pressed to the surface of the paper towels to remove the remaining liquid from the wells.

17. Drag the multichannel pipettor to the pipette tip dispenser to insert the tips.

18. Drag the multichannel pipettor to the developing buffer bottle to draw the developing buffer into the tips. The developing buffer contains the conjugated secondary antibody.

19. Drag the multichannel pipettor to the microtiter plate to dispense the solution into the wells. The tips will automatically be removed and disposed of in the biohazardous waste disposal.

20. Set the timer to 1 hour and then click **Start** to start the timer and allow the conjugated secondary antibody to bind to the primary antibody if it is present in the sample.

21. Drag the washing buffer squeeze bottle to the microtiter plate to remove any nonspecific binding that occurred.

22. Drag the microtiter plate to the sink to dump the contents of the tray into the sink.

23. Drag the microtiter plate to the paper towels. The plate will be pressed to the surface of the paper towels to remove the remaining liquid from the wells.

24. Drag the multichannel pipettor to the pipette tip dispenser to insert the tips.

25. Drag the multichannel pipettor to the substrate bottle to draw the substrate into the tips.

26. Drag the multichannel pipettor to the microtiter plate to dispense the solution into the wells. The tips will automatically be removed and disposed of in the biohazardous waste disposal.

27. An enlargement of the wells will appear. The development will progress over time. To determine the optical density for each sample (the samples are in the first row, from top to bottom, of the microtiter plate):

- Click the well and the optical density will appear in the window of the microtiter plate reader.
- Click **Record Data** to display your results in the grid (and record your results in Chart 3).

CHART 3	Indirect ELISA Results	
Sample	Optical density	HIV test result
Patient A		
Patient B		
Patient C		
Positive control		
Negative control		

28. You will now indicate whether the result for each sample is negative, indeterminate, or positive for HIV.

- A result of <0.300 is read as negative for HIV-1.
- A result of 0.300–0.499 is read as indeterminate (need to retest).
- A result of >0.500 is read as positive for HIV-1.

Click the row for the sample in the grid and then click the − button, **IND,** or the + button above the HIV test result column to indicate whether the result for the sample is positive, indeterminate, or negative for HIV. Repeat this step for all five samples. Record your results in Chart 3.

After you complete the Experiment, take the online **Post-lab Quiz** for Activity 3.

Activity Questions

1. Describe how you can tell that this test is the indirect ELISA rather than the direct ELISA.

_____

_____

2. Describe what the secondary antibody binds to in this activity and why.

_____

_____

3. Define *seroconversion*. How can you tell that a sample has seroconverted?

_____

_____ ▬

Western Blotting Technique

OBJECTIVES

1. To compare the Western blotting technique to the ELISA.
2. To observe the use of the Western blotting technique to test for HIV.
3. To distinguish between antigens and antibodies.

Introduction

Southern blotting was developed by Ed Southern in 1975 to identify DNA. A variation of this technique, developed to identify RNA, was named Northern blotting, thus continuing the directional theme. Western blotting, another variation that identifies proteins, is named by the same convention.

Western blotting uses an electrical current to separate proteins on the basis of their size and charge. This technique uses **gel electrophoresis** to separate the proteins in a gel matrix. Because the resulting gel is fragile and would be difficult to use in further tests, the proteins are then transferred to a **nitrocellulose membrane.** The original Western blotting technique used blotting (diffusion) to transfer the proteins, but electricity is also used now for the transfer of the proteins to nitrocellulose strips. These strips are commercially available, eliminating the need for the electrophoresis and transfer equipment. In this activity you will begin the procedure after the HIV (human immunodeficiency virus) antigens have already been transferred to nitrocellulose and cut into strips.

Western blotting is also known as **immunoblotting** because the proteins that are transferred, or blotted, onto the membrane are later treated with antibodies—the same procedure used in the **indirect enzyme-linked immunosorbent assay (ELISA).** The ELISA is considered enzyme linked because an enzyme is chemically linked to an antibody in both the direct and indirect versions of the test. Immunosorbent refers to the fact that either antigens or antibodies are being adsorbed (stuck) to plastic. If the test is designed to detect an antigen or antigens, it is a **direct ELISA** because it is directly looking for the foreign substance. An **indirect ELISA** is designed to detect antibodies that the patient has made against the antigen.

Similar to the secondary antibodies used in the indirect ELISA technique, the secondary antibodies in the Western blot have an enzyme attached to them, allowing for the use of color to detect a particular protein. The secondary antibody binds to the constant region of the primary antibody found in the patient's sample. The main difference between these techniques is that the ELISA technique uses a well that corresponds to a mixture of antigens, and the Western blot has a discrete protein band that represents the specific antigen that the antibody is recognizing. Like HIV, Lyme disease can also be detected with the Western blot technique.

The initial test for HIV is the ELISA, which is less expensive and easier to perform than the Western blot. The Western blot is used as a confirmatory test after a positive ELISA because the ELISA is prone to false-positive results. The bands from a positive Western blot are from antibodies binding to specific proteins and glycoproteins from the human immunodeficiency virus. A positive result from the Western blot is determined by the presence of particular protein bands (view Table 12.1).

> **EQUIPMENT USED** The following equipment will be depicted on-screen: washing buffer; developing buffer—secondary antibody conjugated with an enzyme; substrate solution; five samples in the samples cabinet: patient A, patient B, patient C, positive control, and negative control; rocking apparatus; nitrocellulose strips; troughs; tray; biohazardous waste disposal.

Experiment Instructions

Go to the home page in the PhysioEx software and click **Exercise 12: Serological Testing.** Click **Activity 4: Western Blotting Technique,** and take the online **Pre-lab Quiz** for Activity 4.

After you take the online Pre-lab Quiz, click the **Experiment** tab and begin the experiment. The experiment instructions are reprinted here for your reference. The opening screen for the experiment is shown below.

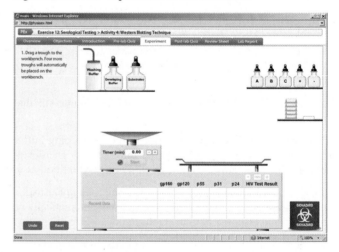

1. Drag a trough to the tray on the workbench. Four more troughs will automatically be placed on the tray.

2. Click the stack of nitrocellulose strips to place a nitrocellulose strip in each trough.

3. Drag the dropper cap of the patient A sample bottle to the first trough to dispense the antiserum from patient A to the nitrocellulose strip. A drop of antiserum for each patient will be dispensed into a separate trough.

4. Drag the tray holding the five troughs to the rocking apparatus.

5. Set the timer to 60 minutes by clicking the **+** button beside the timer display. Click **Start** to gently rock the samples and allow the antibodies to react with the antigens bound to the nitrocellulose. The tray will automatically be returned to the workbench and each trough will be drained into the biohazardous waste disposal when the time is complete. The simulation compresses the 60-minute time period into 10 seconds of real time.

6. Drag the washing buffer squirt bottle to the first trough to dispense washing buffer in each trough. Each trough will be automatically drained into the biohazardous waste container. The washing step removes any nonspecific binding of antibodies that occurred.

7. Drag the dropper cap of the developing buffer bottle to the first trough to dispense developing buffer to each trough.

8. Drag the tray holding the five troughs to the rocking apparatus.

9. Set the timer to 60 minutes by clicking the + button beside the timer display. Click **Start** to gently rock the samples and allow the antibodies to react with the antibodies bound to the nitrocellulose. The tray will automatically be returned to the workbench and each trough will be drained into the biohazardous waste disposal when the time is complete.

10. Drag the washing buffer squirt bottle to the first trough to add washing buffer to each trough. Each trough will be automatically drained into the biohazardous waste container. The washing step removes any nonspecific binding of secondary conjugated antibodies. Excess secondary conjugated antibodies could react erroneously with the substrate and give a false-positive result.

11. Drag the dropper cap of the substrates bottle to the first trough to dispense the substrates (tetramethyl benzidine and hydrogen peroxide) into each trough. The substrates are the chemicals that are being changed by the enzyme that is linked to the antibody.

12. Drag the tray holding the five troughs to the rocking apparatus.

13. Set the timer to 10 minutes by clicking the + button beside the timer display. Click **Start** to gently rock the samples and allow the enzyme to react with the substrates. The tray will automatically be returned to the workbench when the time is complete. The simulation compresses the 10-minute time period into 10 seconds of real time.

14. To determine the antigens present for each sample:

- Click the nitrocellulose strip inside the trough to visualize the results.

- Click **Record Data** to display your results in the grid (and record your results in Chart 4). The bands present on the nitrocellulose strip represent the antibodies present in the sample that have reacted with the antigens (bands) on the strip (view Table 12.1).

Repeat this step for all five samples.

TABLE 12.1	HIV Antigens
Abbreviation	Description
gp160	Glycoprotein 160, a viral envelope precursor
gp120	Glycoprotein 120, a viral envelope protein that binds to CD4
p55	A precursor to the viral core protein p24
gp41	A final envelope glycoprotein
p31	Reverse transcriptase
p24	A viral core protein

15. You will now indicate whether the result for each sample is negative, indeterminate, or positive for HIV. The criteria for reporting a positive result varies slightly from agency to agency. The Centers for Disease Control and Prevention recommend the following criteria:

- If no bands are present, the result is negative.

- If bands are present but they do not match the criteria for a positive result, the result is indeterminate. Patients that are deemed indeterminate after multiple tests should be monitored and tested again at a later date.

- If either p31 or p24 is present *and* gp160 or gp120 is present, the result is positive. Click the row for the sample in the grid and then click the − button, **IND,** or the + button above the HIV test result column to indicate whether the result for the sample is positive, indeterminate, or negative for HIV.

Repeat this step for all five samples. Record your results in Chart 4.

After you complete the experiment, take the online **Post-lab Quiz** for Activity 4.

CHART 4	Western Blot Results					
Sample	gp160	gp120	p55	p31	p24	HIV test result
Patient A						
Patient B						
Patient C						
Positive control						
Negative control						

Activity Questions

1. Describe how gel electrophoresis is used to separate proteins.

_____

_____

2. In a patient sample that is positive for HIV, would antibodies or antigens be present when using the Western blot technique? How do you know?

_____

_____

NAME_____

LAB TIME/DATE_____

Serological Testing

ACTIVITY 1 Using Direct Fluorescent Antibody Technique to Test for Chlamydia

1. Describe the importance of the washing steps in the direct antibody fluorescence test. _____

   _____

   _____

2. Explain where the epitope (antigenic determinant) is located. _____

   _____

   _____

3. Describe how a positive result is detected in this serological test. _____

   _____

   _____

4. How would the results be affected if a negative control gave a positive result? _____

   _____

   _____

ACTIVITY 2 Comparing Samples with Ouchterlony Double Diffusion

1. Describe how you were able to determine what antigen is in the unknown well. _____

   _____

   _____

2. Why does the precipitin line form? _____

   _____

   _____

3. Did you think human serum albumin and bovine serum albumin would have epitopes in common? How well did the results

compare with your prediction? _____

_____

_____

ACTIVITY 3 Indirect Enzyme-Linked Immunosorbent Assay (ELISA)

1. Describe how the direct and indirect ELISA are different. _____

_____

_____

2. Discuss why a patient might test indeterminate. _____

_____

_____

3. How would your results have been affected if your negative control had given an indeterminate result? _____

_____

_____

4. Briefly describe the basic structure of antibodies. _____

_____

_____

ACTIVITY 4 Western Blotting Technique

1. Describe why the HIV Western blot is a more specific test than the indirect ELISA for HIV. _____

_____

_____

2. Explain the procedure for a patient with an indeterminate HIV Western blot result. _____

_____

_____

3. Briefly describe how the nitrocellulose strips were prepared before the patient samples were added to them. _____

_____

_____

4. Describe the importance of the washing steps in the procedure. _____

_____

_____

Index

NOTE: Page numbers in **boldface** indicate a definition. A *t* following a page number indicates tabular material and an *f* indicates an illustration.

Abdominal/abdominal wall muscles, in
  breathing, **104**, **105**
ABO antigens/blood group, 160, **165**, 166*t*
  inheritance of, 165
Absolute refractory period, **43**, **92**
  cardiac, **92**
  neuron, **42–43**
Acetylcholine, **96**
  in muscle contraction, 16
Acid, **147**
  strong, **147**
  weak, **147**
Acid-base balance, **147–157**
  renal system in, **148**, 150–152
  respiratory system in, 148–150, **148**,
    152–154
Acidosis, **148**, **149**
  metabolic, **152**
  respiratory, 149, 151
  respiratory and renal responses to, 148, 149,
    150–152, 152–154
ACTH (adrenocorticotropic hormone),
  67–68, 67*t*
Action potential, **33**, **39**
  cardiac, **92**, 92*f*
  in neurons, **33**, **39**, 48–50
    chemical synaptic transmission/
      neurotransmitter release
      and, **47–48**
    conduction velocity and, 45–47, **46**
    refractory periods (absolute and relative)
      and, 42–43, **43**
    stimulus intensity coding and, 44–45
    threshold and, 39–40
    voltage-gated sodium channels and, **40–42**
Active force, in muscle contraction, **24–25**
Active processes, **1**, **2**, **11**
  active transport, **2**, 11–12
    secondary, **138**
  vesicular transport, **2**
Active site, **121**
Active transport, **2**, 11–12
  secondary, **138**
Adaptation, 44
Addison's disease, **67**, 67*t*
Adequate stimulus, 37, 38
ADH. *See* Antidiuretic hormone
Adrenal cortex, 67
Adrenal glands, 58*f*
Adrenaline. *See* Epinephrine
Adrenal insufficiency, 67, 67*t*
Adrenergic modifiers, **96**
Adrenocorticotropic hormone (ACTH),
  67–68, 67*t*
Afferent arteriole, **129**, **131**, **133**, **135**
  radius of, glomerular filtration affected by,
    130–133, **135–137**
A fibers, conduction velocity of, 46, 46–47
Afterload, **82**, **84**
Agglutinins, blood typing and, **165**
Agglutinogens, blood typing and, **160**, **165**
Agonist, **96**
Alcohol ingestion, metabolic acidosis and, 152

Aldosterone, **140**
  urine formation and, **140–141**
Alkali ingestion, metabolic alkalosis and, 152
Alkalosis, **148**
  metabolic, **152**
  respiratory, **148**, 150
  respiratory and renal responses to, 148,
    150–152, 152–154
All-or-none phenomenon, action potential
  as, 44, 46
Alveoli, respiratory, **103**
Amino acids, **123**
Anabolism, 58
Anemia, **160**
  blood viscosity and, 76
  hemoglobin and, 164
Angiotensin converting enzyme, 140
Antacid ingestion, metabolic alkalosis and, 152
Antagonist, **96**
Anti-A antibodies, 166, 166*t*
Anti-B antibodies, 166, 166*t*
Antibodies, 175–176, 176*f*
  blood typing and, 166, 166*t*
  constant region of, 180
  fluorescent, in direct fluorescent antibody
    technique, 176–178
  variable region of, 180
Antidiuretic hormone (ADH), **137**, **140**
  urine concentration and, 137–138
  urine formation and, **137**, **140–141**
Antigen(s), **175–176**, 176*f*
  blood typing and, 160, **165**, 166*t*
  Ouchterlony technique in identification
    of, 178–179
Antigen-antibody specificity, 176, 176*f*
Antigenic-binding sites, 176, 176*f*
Antigenic determinants (epitopes), **176**, 176*f*
Antiporter, 11
Aortic valve stenosis, 84
Apical membrane, **138**
Aplastic anemia, 160
Arterioles, in nephron, 129, 131
  radius of, glomerular filtration affected by,
    130–133, 135–137
Arteriosclerosis, **84**
Asthma, **107**, 109
  inhaler medication and, 109
Atelectasis, **110**
Atherosclerosis, **84**, **167**
ATP, for active processes, 1, 2, 11
Atropine, heart rate and, 96–97
Autonomic nervous system (ANS)
  in heart regulation, 93, 96
  parasympathetic division of, **93**
  sympathetic division of, **93**
Autorhythmicity, **91**
Axon(s), **34**, 34*f*, 39, 47
Axon hillock, **39**
Axon terminal, 34, 34*f*, 47

BAPNA, **123**, 124
Basal metabolic rate (BMR). *See also*
  Metabolic rate/metabolism
  determining, 59–60
  thyroid hormone affecting, 60
Base (chemical), **147**
  strong, **147**
  weak, **147**

Basolateral membrane, **138**
Benedict's assay, 119, 120, 121, 122
B fibers, conduction velocity of, 46, 47
Bicarbonate, in acid-base balance, 147, 148,
  149, 151, 152
Bicarbonate buffer system, 148
Bicarbonate ingestion, metabolic alkalosis
  and, 152
Bile, 125
Bile salts, 125
Biological action, hormone-receptor complex
  exerting, **57**
Blood
  analysis of, 159–174. *See also specific test*
  pH of, 147
Blood cells (formed elements), blood viscosity
  and, 76
Blood cholesterol, **159**, **167–169**
Blood flow, **74**
  pressure and resistance and, 73–74, 74,
    74–86
    blood pressure and, 78–80
    compensation/cardiovascular pathology
      and, 84
    vessel length and, 74, 77–78
    vessel radius and, 74, **74–76**
    viscosity and, 74, 76–77
Blood pressure (BP)
  blood flow and, 78–80
  glomerular filtration and, 133–135, **135–137**
  renal response to changes in, 135–137
Blood type, 159, 160, 165–167, 166*t*
  inheritance of, 165
Blood vessel length, blood flow and, 74, 77
Blood vessel radius, **74**
  blood flow and, 74, **74–76**
  glomerular filtration and, 130–133, 135–137
  pump activity and, 80–81
Blood vessel resistance, **73**, **74**
  blood flow and, 73–74, 74, 74–86
    compensation/cardiovascular pathology
      and, 84
  blood pressure and, 73–74, 74
Blood viscosity, 74, 76
  blood flow and, 74, 76–77
Body temperature regulation, 95
Bone density, loss of in osteoporosis, 65
Bowman's (glomerular) capsule, 129, 130*f*,
  131, **133**, **135**
Breathing (ventilation), **103**, 104, 104*f*
  acid-base balance and, 148, 148–150,
    152–154
  frequency of, **104**
  mechanics of, 103–115
  muscles of, **104**, **105**
Buffer system(s), 148
  chemical, 148
  renal, **148**
  respiratory, **148**
Buffy coat, 160

Calcitonin, 65, 66
Calcium
  autorhythmicity and, 91
  heart rate and, 97–99, 98*t*
  neurotransmitter release and, 47
Calcium channel blockers, 97, 98
cAMP, 57–58

Capillaries
  glomerular, **131**, **133**
  peritubular, **130**, **137**
Carbohydrate (starch) digestion, 119–121
  by salivary amylase, **119**–121
    substrate specificity and, 121–123
Carbon dioxide
  acid-base balance and, 148, 149, 151, 152
  partial pressure of, 104, 148, 151
Carbonic acid, in acid-base balance, 148, 149, 151, 152
Cardiac action potential, **92**, 92*f*
  ion movement and, 98*t*
Cardiac cycle, 80
Cardiac muscle
  contraction of, 82, 91
  refractory period of, **92**–93
Cardiac output, **74**, **80**, **82**
Cardiac pacemaker, 91, **94**
Cardiac valves, 84
Cardiovascular compensation
  pathological conditions and, 84–86
  vessel radius affecting pump activity and, 80–81
Cardiovascular dynamics, 73–89
Cardiovascular physiology, 91–102
Carrier proteins, **138**
  active transport and, 2, 11
  facilitated diffusion and, 1, 4
  glucose transport/reabsorption and, 4–6, **138**–140
Catabolism, 58
Catalysts, 118
Catecholamine hormones, 57
C cells, 65
Cell(s), permeability/transport mechanisms of, 1–14
Cell body, neuron, 33, 34*f*
Cellulose, **121**
Cellulose digestion, 121–123
C fibers, conduction velocity of, 46, 47
Chemical buffers, 148
Chemical modifiers, heart rate and, 96–97
Chemical neurotransmitters, **33**
Chemical synapses, 33, **47**–48, 48
Chief cells, **123**
Chlamydia/*Chlamydia trachomatis,* **176**
  direct fluorescent antibody technique in testing for, **176**–178
Cholesterol, 159, 160, **167**
  blood concentration of, **159**, **167**–169
Cholinergic modifiers, **96**
Chronotropic modifiers, **98**
Collecting ducts, 130, 130*f*, **137**
Comparative spirometry, 107–110
Compensation, cardiovascular
  pathological conditions and, 84–86
  vessel radius affecting pump activity and, 80–81
Compensatory pause, 93
Complete (fused) tetanus, **22**, **23**, 92
Concentration gradient, **2**
  in active transport, 2, 11
  in diffusion, 1, **2**, 4
  in osmosis, 6
  urine concentration and, 137–138
Concentric contraction, isotonic, **26**
Conductance, **34**
Conduction, action potential, 39, 46
Conduction velocity, 45–47, **46**
Constant region (antibody), 180
Constipation, metabolic alkalosis and, 152

Contractility (heart), **82**
  compensation and, 84
Contraction, skeletal muscle, 15–32. *See also* Muscle contraction
Contraction phase, of muscle twitch, **16**, 18
Controls, 119, 123, 176
Converting enzyme (angiotensin), 140
Corticotropin-releasing hormone (CRH), 67
Cortisol, 67–68, 67*t*
Coupled transport, 11
Cranial nerve X (vagus nerve), heart affected by stimulation of, 93–94
CRH. *See* Corticotropin-releasing hormone
Cushing's disease, 67, 67*t*
Cushing's syndrome, 67, 67*t*
Cyclic adenosine monophosphate (cAMP), 57–58

Dendrite(s), **33**, 34*f*
Depolarization, **37**
  cardiac, 92
    ion movement and, 98*t*
  muscle cell, 16
  neuron, 37, 39
Diabetes mellitus, 62–65
  fasting plasma glucose levels and, 63, 64
Dialysis, **1**, 2–4. *See also* Simple diffusion
Diaphragm, **104**, **105**
Diarrhea, metabolic acidosis and, 152
Diastole, 80
Differential (selective) permeability, **1**, 2, 6
Diffusion, **1**, **2**
  facilitated, 1, **4**–6, **138**
  simple, 1, **2**–4
Digestion, **117**–128
  chemical (enzymatic action), 118, 118*f*
Digestive system, **117**, 118*f*
Digitalis, heart rate and, 96–97
Direct enzyme-linked immunosorbent assay (ELISA), **180**, **182**
Direct fluorescent antibody technique, 176–178
Distal convoluted tubule, **129**, 130*f*
Double diffusion, Ouchterlony, 178–179
Dual X-ray absorptiometry (DXA), 65
DXA. *See* Dual X-ray absorptiometry

EDV. *See* End diastolic volume
Efferent arteriole, **129**, **131**, **135**
  radius of, glomerular filtration affected by, 130–133, **135**–137
Electrical stimulation, muscle contraction and, **15**, 16
Electrophoresis, gel, **182**
Elementary body (chlamydia), **176**
ELISA. *See* Enzyme-linked immunosorbent assay
Emphysema, **107**, 108
End diastolic volume (EDV), **80**, **82**
Endocrine system/glands, **57**, 58*f*
  physiology of, **57**–72
End-plate potential, **16**
End systolic volume (ESV), **80**, **82**
Energy
  for active processes, 1, 2, 11
  kinetic, **1**, 2
Enzyme(s), **118**
  in digestion, 118, 118*f*
  substrate specificity of, amylase, 121–123
Enzyme assay, **119**, 121
Enzyme-linked immunosorbent assay (ELISA), **180**

direct, **180**, **182**
  indirect, 179–182, **180**, **182**
Enzyme substrates, **118**, **121**, **123**
Epinephrine, **96**
  heart rate and, **96**–97
Epitopes (antigenic determinants), **176**, 176*f*
Equilibrium (solution), **1**
ERV. *See* Expiratory reserve volume
Erythrocyte(s) (red blood cells/RBCs)
  in hematocrit, 159, 160
  settling of (ESR), 159, **159**–160, **162**–163
Erythrocyte (red blood cell) antigens, blood typing and, 160, **165**, 166*t*
Erythrocyte sedimentation rate (ESR), **159**, **159**–160, **162**–163
Erythropoietin, 159
ESR. *See* Erythrocyte sedimentation rate
Estrogen(s), **65**
  bone density and, 65, 66
Estrogen (hormone) replacement therapy, 65–67
ESV. *See* End systolic volume
Excitation-contraction coupling, **16**
Excitatory postsynaptic potential, 49
Excretion, by kidneys, 129, 133
Exercise
  breathing during, 104, **107**, 109
  metabolic acidosis and, 152
Expiration, **104**, 105
Expiratory reserve volume (ERV), **105**
External intercostal muscles, **104**, **105**
External respiration, 104*f*
Extrasystoles, **93**

Facilitated diffusion, 1, **4**–6, **138**
"False negative," 176
"False positive," 176
Fasting plasma glucose, 63, 64–65
Fat digestion, 125–126
Fatigue, muscle, **23**–24
Fatty acids, 125
Feedback mechanisms
  negative, **58**
    thyroid hormone secretion and, 59
  positive, 58
FEV$_1$ (forced expiratory volume), **105**
Filtrate, 9, **131**, **133**
Filtrate pressure, 133
Filtration, **1**, **1**–2, **9**–11
  glomerular, **129**, **130**
    arteriole radius affecting, 130–133, **135**–137
    pressure affecting, 133–135, **135**–137
Fluorescent antibody technique, direct, 176–178
Follicle-stimulating hormone (FSH), **65**
Food, digestion of, **117**–128
  chemical (enzymatic action), 118, 118*f*
Force, muscle, **15**, **18**
Forced expiratory volume (FEV$_1$), **105**
Forced vital capacity (FVC), **105**
Force transducer, **15**
Formed elements of blood (blood cells), blood viscosity and, 76
FPG. *See* Fasting plasma glucose
Frank-Starling law of the heart, 82
Frequency (stimulus), **20**, **22**, **23**
  muscle contraction affected by, **20**–21
  tetanus and, **22**, **23**, 92
Frequency of breathing, **104**
Frog, cardiovascular physiology in, 91–102
  poikilothermia and, **95**
FSH. *See* Follicle-stimulating hormone
Fused (complete) tetanus, **22**, **23**, 92
FVC (forced vital capacity), **105**

Gas exchange, 103, 104, 104f
Gastric lipase, **125**
Gastrointestinal (GI) system, 117, 118f
Gastrointestinal (GI) tract (alimentary canal), 117
Gated channels, 34
  voltage-gated potassium channels, refractory periods and, 42
  voltage-gated sodium channels, **40**–42
  refractory periods and, 42
Gel electrophoresis, **182**
Glia, myelination and, 46
Glomerular capillaries, **131**, **133**
Glomerular capillary pressure, 131, 133, **135**
  glomerular filtration affected by, 133–135, 135–137
Glomerular (Bowman's) capsule, 129, 130f, **131**, **133**, **135**
Glomerular filtration, **129**, **130**
  arteriole radius affecting, 130–133, 135–137
  pressure affecting, 133–135, **135**–137
Glomerular filtration rate, **131**, **133**, 135. See also Glomerular filtration
Glomerulus (glomeruli), 129, **130**–131, 130f, **133**, **135**
Glucagon, **62**
Glucose, **62**–65
  plasma, **62**–65
    fasting levels of, 63, 64–65
  transport/reabsorption of, carrier proteins and, 4–6, **138**–140
Glucose carrier proteins, 4–6, **138**–140. See also Carrier proteins
Glucose standard curve, **63**
  developing, 63–64
Goiter, **58**
gp41, 183t
gp120, 183, 183t
gp160, 183, 183t
Graded depolarization, 16
Graded potentials, 37
Gradient, 1

Hb. See Hemoglobin
Heart, 80
  contractility of, **82**
  epinephrine affecting, **96**
  Frank-Starling law of, 82
  nervous stimulation of, 93–94
  physiology of, 91–102
  pumping activity of, 80
    vessel radius affecting, 80–81
  valves of, 84
Heart rate
  in cardiac output, 80, 82
  chemical modifiers of, 96–97
  ions affecting, 97–99, 98t
  temperature affecting, 94–96
Heat receptors, 38
Hematocrit, **74**, **159**, **160**–162
Heme, **164**
Hemoglobin, **159**, **160**, 163–165, **164**
Hemoglobinometer, **164**
Henle, loop of, **129**, 130f
Hertz (Hz), 45
HIV
  ELISA in testing for, 180–182, 182
  Western blotting in testing for, 182–184
HIV antigens, 182, 183t
Homeostasis, **57**
Homeothermic animals, humans as, 58, **95**
Hormone(s), **57**

  metabolism and, **58**–62
  tropic, 59, 67
  urine formation and, 140–141
Hormone replacement therapy, 65–67
Human immunodeficiency virus (HIV)
  ELISA in testing for, 180–182, 182
  Western blotting in testing for, 182–184
Human immunodeficiency virus (HIV) antigens, 182, 183t
Hydrolases, **118**
Hydrolysis, **123**
  of fat, by lipase, 125–126
  of proteins, by pepsin, **123**–124
  of starch, by salivary amylase, 119–121
Hypercholesterolemia, **168**
Hypercortisolism (Cushing's syndrome/Cushing's disease), 67, 67t
Hyperpolarization, 39
Hyperthermia, **95**
Hypertonic solution, **7**
Hyperventilation, acid-base balance and, 148–149
Hypocholesterolemia, **168**
Hypocortisolism, 67
Hypophysectomy, 59
Hypophysis (pituitary gland), 58f
  hypothalamus relationship and, 59
  thyroxine production regulated by, 58
Hypopituitarism, 67t
Hypothalamic-pituitary portal system, **59**
Hypothalamus, 58, 58f
  pituitary relationship and, 59
  thyroxine/TSH production and, **58**, 59
Hypothermia, **95**
Hypotonic solution, **7**
Hypoventilation, acid-base balance and, 149
Hz. See Hertz

Iatrogenic, 67
Iatrogenic Cushing's syndrome, 67, 67t
Identity (antigen), **178**
  partial, **178**
IKI assay, 119, 120, 121, 122
Immunoblotting (Western blotting), **182**–184
Indirect enzyme-linked immunosorbent assay (ELISA), 179–182, **180**, **182**
Inhaler (asthma), 109
Initial segment, 39
Inotropic modifiers, 98
Inspiration, 104, **104**–105
Inspiratory reserve volume (IRV), **105**
Insulin, **62**–65
Intercostal muscles
  external, **104**, **105**
  internal, **104**, **105**
Internal intercostal muscles, **104**, **105**
Internal respiration, 104f
Interneurons, **33**, 48
Interspike interval, 45
Interstitial spaces, nephron, **137**
Intrapleural pressure, **110**–111
Intrapleural space, **110**
Ion(s), **34**
  heart rate and, 97–99, 98t
Ion channels, **34**
Iron deficiency anemia, 160
IRV. See Inspiratory reserve volume
ISI. See Interspike interval
Isometric contraction, 18, **24**–26, 26
Isometric length-tension relationship, **25**
Isotonic concentric contraction, 26

Isotonic contraction, **26**–27
Isotonic solution, **7**

Ketoacidosis, **152**
Kidney(s), **129**
  in acid-base balance, 148, 150–152
  physiology of, **129**–145
Kinetic energy, **1**, 2

Lamellar (Pacinian) corpuscle, 37
Laminar flow, **74**
Latent period
  in isotonic concentric contraction, 26
  of muscle twitch, 16, 18
LDL. See Low-density lipoprotein
Leak channels, 34
Length-tension relationship
  cardiac muscle, **82**
  skeletal muscle, 24–26
    isometric, **25**
Lidocaine, 40, 41
Light receptors, 38
Lingual lipase, **125**
Lipase, **125**
  in fat digestion, **125**–126
  gastric, **125**
  lingual, **125**
  pancreatic, **125**
Lipids, 125
Lipoproteins, **160**, **167**
Liver, 118f
Load, skeletal muscle affected by, **26**–27
Load-velocity relationship, 26–27
Local potentials, 37
Loop of Henle, **129**, 130f
Low-density lipoprotein (LDL), **167**–168
Lung(s), mechanics of respiration and, 103–115
Lung capacity, total, **105**

Maltose/maltotriose, starch hydrolyzed to, 119–121, 121–123
Maximal shortening velocity, 26
Maximal stimulus, muscle contraction and, **18**
Maximal tetanic tension, **22**, **23**
Maximal voltage, muscle contraction and, **18**
Membrane(s), diffusion through, **1**, 2
  facilitated diffusion, **1**, 4–6
  simple diffusion, **1**, 2–4
Membrane pores, in filtration, 9
Membrane potential, **34**
  resting, **34**
    cardiac cell, 97
    ion movement and, 98t
    neuron, **34**–37
Membrane proteins, neural, 34, 37
Mercury, millimeters of, osmotic pressure measured in, 7
Metabolic acidosis/metabolic alkalosis, **152**
  renal and respiratory response to, 152–154
Metabolic rate/metabolism, **58**
  acidosis/alkalosis and, 152
  computing, 59–60
  thyroid hormone and, **58**–62
Millivolt(s), **34**
Minor calyces, 130
Minute ventilation, **104**
Molecular weight cutoff (MWCO), 2, 9
Motor end plate, 16
Motor neurons, 16, 33
Motor unit, **16**, **18**
  recruitment and, **18**
  wave summation and, **20**, 92

Motor unit recruitment, **18**
Muscle contraction, **15**–32
  electrical stimulation and, **15**, 16
  fatigue and, **23**–24
  isometric, 18, **24**–26, 26
  isotonic, **26**–27
  length-tension relationship in, 24–26
    isometric, 25
  muscle twitch and, **16**–18, **18**
  stimulus frequency and, **20**–21, 22
  stimulus voltage and, **18**–20
  tetanus and, **22**–23, 23, 92
  threshold stimulus for, **18**
  treppe and, **20**
Muscle fatigue, **23**–24
Muscle fibers (muscle cells), **15**, 16, 16f
  contraction of, 15–32. *See also* Muscle
    contraction
Muscle tension (muscle force), **15**, 18
Muscle twitch, **16**–18, **18**
Muscular system/muscles, structure of, 15, 16f
mV. *See* Millivolt(s)
MWCO. *See* Molecular weight cutoff
Myelination, **46**
Myometrium, oxytocin affecting, 58

Na⁺-K⁺ pump (sodium-potassium pump),
  11, 34
Negative controls, 119, 123, **176**
Negative feedback mechanisms, **58**
  thyroid hormone secretion and, 59
Negative starch test, 120, 122
Negative sugar test, 120, 122
Nephron(s), **129**, **130**, **135**, **151**
  function of, **129**–146
Nerve(s), **39**
  physiology of, 33–53
Nerve chamber, 39
Nerve conduction velocity, 45–47, **46**
Nerve impulse. *See also* Action potential
  neurophysiology of, 33–53
Nervous system
  autonomic, 93
    in heart regulation, 93, 96
  physiology/neurophysiology of, 33–53
Neuroglia, myelination and, 46
Neuromuscular junction, **16**
Neuron(s), **33**–34, 34f
  physiology/neurophysiology of, **33**–53
Neurophysiology, 33–53
Neurotransmitters, **33**, 47–48
Nitrocellulose membrane, **182**
Nitrogen balance, negative, in diabetes, 62
Non-identity (antigen), **178**
Nonself antigens, **175**
Nonspecific binding, **176**
Norepinephrine/noradrenaline, **96**

Obstructive lung disease, **106**
Odor receptors. *See* Olfactory (odor) receptors
Oils, 125
Olfactory (odor) receptors, 38
  adaptation and, 44
Oligodendrocytes, 46
Optical density, **123**, 124
  developing standard glucose curve and,
    63–64
  measuring fasting plasma glucose and, 64
Oscilloscope, membrane potential studied
  with, 34
Osmosis, **1**, **2**, **6**–8
  tonicity and, 7
Osmotic pressure, **6**–8

Osteoporosis, **65**
Ouchterlony double diffusion, 178–179
Ovariectomy, 65
Ovary, 118f
Oxygen, partial pressure of, 104
Oxygen consumption, thyroid hormone effect
  on metabolism and, 59, 59–60
Oxyhemoglobin, **164**
Oxytocin release, positive feedback
  regulating, 58

p24, 183, 183t
p31, 183, 183t
p55, 183t
Pacemaker, cardiac, 91, **94**, 96
Pacinian (lamellar) corpuscle, 37
Pancreas, 58f, 118f
Pancreatic lipase, **125**
Parasympathetic nervous system, **93**
  in heart regulation, 93, 96
Parathyroid glands, 58f
Partial identity (antigen), 178
Partial pressure of gas, **104**
  carbon dioxide, 104, 148, 151
  oxygen, 104
Passive force, in muscle contraction, 24
Passive processes/transport, **1**
  diffusion, **1**, **2**
    facilitated diffusion, **1**, **4**–6
    simple diffusion, **1**, **2**–4
  filtration, **1**, **1**–2, **9**–11
P_{CO_2} (partial pressure of carbon dioxide), 104,
  148, 151
Pepsin, **123**
  protein digestion by, **123**–124
Peptidase, **121**
Peptide hormones, 57
Peptides, 57, **123**
  digestion of, **123**–124
Peripheral resistance, **73**, 74
  blood flow and, 73–74, 74, 74–86
    compensation/cardiovascular pathology
      and, 84
  blood pressure and, 73–74, 74
Peritubular capillaries, **130**, 137
Permeability (cell), 1–14
Permeability (ion channel), **34**
pH, **125**, 147
  buffering systems in maintenance of, 148
    renal, **148**, 151–152
    respiratory, 148–150, **148**
  lipase activity and, 125
  normal, 147, 148, 151
pH meter, 125
Phosphate buffer system, 148
Physiological buffering systems, 148
Pilocarpine, heart rate and, 96–97
Pineal gland, 58f
Pituitary gland, 58f
  hypothalamus relationship and, 59
  thyroxine production regulated by, 58
Plasma glucose, **62**–65
  fasting, 63, 64–65
Plasma membrane
  diffusion through, **1**, **2**
    facilitated diffusion, **1**, **4**–6
    simple diffusion, **1**, **2**–4
  permeability of, 1–14
Plateau phase, in cardiac action potential,
  92, 92f
  ion movement and, 98t
Pneumothorax, **110**
Poikilothermic animal, frog as, **95**

Polarization
  cardiac, 92, 92f
  neuron, **34**
Polycythemia, **160**
  blood viscosity and, 76
Polypeptide, **123**
Pore(s), membrane, in filtration, 9
Portal vein, **59**
Positive controls, 119, **176**
Positive feedback mechanisms, 58
Positive starch test, 120, 122
Positive sugar test, 120, 122
Postsynaptic potential, **47**, 49
Potassium, heart rate and, 97–99, 98t
Potassium channels, 34
  voltage-gated, refractory periods and, 42
Precipitin line, **178**
Preload, **82**
Pressure
  blood flow and, 78–80
  glomerular filtration and, 133–135, **135**–137
  osmosis affecting, 6
Pressure gradient
  blood flow and, **73**, 79
  filtration and, 1–2, 9
Pressure receptors, 38
Propagation, action potential, **39**, 46
Propylthiouracil, metabolic rate affected by, 61
Protein(s), **123**
Protein buffer system, 148
Protein digestion, **123**–124
Proximal convoluted tubule, **129**, 130f
  glucose reabsorption in, 138–140
Pulmonary function tests, **105**, 106
  forced expiratory volume (FEV₁), **105**
  forced vital capacity (FVC), **105**
Pulmonary minute ventilation, **104**
Pump activity (heart), 80
  stroke volume affecting, **82**–83
  vessel radius affecting, 80–81

RBCs. *See* Red blood cell(s)
Reabsorption, tubular, **130**, **137**, **138**
  glucose carrier proteins and, **138**–140
  urine concentration and, 137–138
Rebreathing, acid-base balance and, 149–150
Receptor(s), hormone, **57**
Receptor potential, **33**, **37**–39
Receptor proteins, 37
Recruitment (motor unit), **18**
Red blood cell(s) (RBCs/erythrocytes)
  in hematocrit, 159, 160
  settling of (ESR), 159, **159**–160, **162**–163
Red blood cell (erythrocyte) antigens, blood
  typing and, 160, 165, 166t
Red blood cell (erythrocyte) sedimentation rate
  (ESR), 159, **159**–160, **162**–163
Refractory period
  cardiac muscle, **92**–93
  neuron, 42–43, **43**
Relative refractory period, **92**
  cardiac, **92**
  neuron, 42–43
Relaxation phase, of muscle twitch, 16, 18
Renal buffering system, **148**
  in respiratory acidosis/alkalosis, 148, 150–152
Renal calyces, 130
Renal compensation, **151**
Renal corpuscle, **129**, **130**, 130f, 131
Renal physiology, 129–145
Renal tubule, **129**, **130**, 130f, **133**, **135**
  glucose reabsorption and, 138–140
Renal tubule lumen, **137**

Renal tubule pressure, 133
  glomerular filtration affected by, 133–135,
    135–137
Renin-angiotensin system, 140
Repolarization
  cardiac, 92, 92*f*
    ion movement and, 98*t*
  neuron, 42
Reserve volume
  expiratory, **105**
  inspiratory, **105**
Residual volume (RV), **105**
Resistance, vascular, **73**, 74
  blood flow and, 73–74, 74, 74–86
    compensation/cardiovascular pathology
      and, 84
  blood pressure and, 73–74, 74
Respiration, **103**
  mechanics of, 103–115
Respiratory acidosis, 149, 151
  renal response to, 150–152
  respiratory response to, 149, 150–152
Respiratory alkalosis, **148**, 151
  renal response to, 148, 150–152
  respiratory response to, 148
Respiratory buffering system, **148**
  in metabolic acidosis/metabolic alkalosis,
    152–154
  in respiratory acidosis/respiratory alkalosis,
    148, 149, 150–152
Respiratory capacities, 105
Respiratory system
  in acid-base balance, 148–150, **148**, 152–154
  mechanics of, 103–115
Respiratory volumes, 104–107, 105
  spirometry in study of, 107–110
Resting membrane potential, **34**
  cardiac cell, 97
    ion movement and, 98*t*
  neuron, **34**–37
Rest periods, muscle fatigue and, **23**
Restrictive lung disease, **106**
Reticulate body (chlamydia), **176**
Rh antigens/blood group/factor, **165**, 166
Ringer's solution/irrigation, 95
Rouleaux formation, 160, **162**
RV. *See* Residual volume

Salicylate poisoning, metabolic acidosis and, 152
Salivary amylase, **119**, **121**
  substrate specificity of, 121–123
Salivary glands, 118*f*
SA node. *See* Sinoatrial (SA) node
Sarcolemma, **16**
Schwann cells, 46
Secondary active transport, **138**
Sedimentation rate, 159, **159**–160, **162**–163
Selective (differential) permeability, **1**, 2, 6
Self antigens, 175
Semipermeable membrane, 2, 6
Sensitivity, hormone receptor, **57**
Sensory neurons, 33, 37, 39
Sensory receptors, **37**
  adaptation and, 44
Sensory transduction, **37**
Seroconversion, **180**
Serological testing/serology, **175**–186. *See also*
  specific test
Serum, **175**
Shortening velocity, 26
Sickle cell anemia, 160
Simple diffusion, **1**, **2**–4
Sinoatrial (SA) node, **93**–94, **96**

Skeletal muscles
  contraction of, 15–32. *See also* Muscle
    contraction
  length-tension relationship in, 24–26
    isometric, **25**
  load affecting, **26**–27
  physiology of, 15–32
  structure of, 15, 16*f*
Small intestine, 118*f*
Sodium
  autorhythmicity and, 91
  heart rate and, 97–99, 98*t*
  neuron membrane potential and, 34
Sodium channels, 34
  voltage-gated, **40**–42
    refractory periods and, 42
Sodium-potassium pump, 11, 34
Solute gradient, urine concentration and,
  137–138
Solute particles, 2
Solute pumps, **11**
Southern blotting, 182
Specificity (substrate), amylase, 121–123
Spectrophotometer, **123**, 124
  developing standard glucose curve and, 63–64
  measuring fasting plasma glucose and, 64
  in pepsin activity analysis, **123**, 124
Spirometry/spirometer, 107–110
Starch, 121
Starch digestion, 119–121
  by salivary amylase, **119**–121
    substrate specificity and, 121–123
Starch test, positive/negative, 120, 122
Starling forces, **133**, **135**
Steroid hormones, 58
Stimulus frequency, **20**, **22**, **23**
  coding stimulus intensity as, 44–45
  muscle contraction affected by, **20**–21
  tetanus and, **22**, **23**, 92
Stimulus intensity
  coding for, 44–45
  graded/local potentials and, 37
Stimulus voltage, **18**
  muscle contraction affected by, **18**–20
Stomach glands, 118*f*, 123
Stroke volume (SV), **80**, **82**
  in cardiac output, 80, 82
  pump activity affected by, **82**–83
Strong acid, **147**
Strong base, **147**
Substrate(s), enzyme, **118**, **121**, **123**
Substrate specificity, amylase, 121–123
Sugar test, positive/negative, 120, 122
Surface tension, **110**
Surfactant, **110**–111
SV. *See* Stroke volume
Sympathetic nervous system, **93**
  in heart regulation, 93, 96
Symporter, 11
Synapses, chemical, **33**, **47**–48, 48
Synaptic gap/synaptic cleft, 47
Synaptic potentials, **33**–34, 49
Synaptic vesicles, **47**
Systole, **80**

T_4 (thyroxine/thyroid hormone), 58, **58**
  metabolism/metabolic rate and, **58**–62
Target cells, **57**
Temperature, heart rate affected by, 94–96
Tendon(s), **15**, 16*f*
Tension (muscle tension/force), **15**, **18**
Testis, 118*f*
Tetanus, **22**–23, 23, 92

Tetraiodothyronine. *See* Thyroxine
Tetrodotoxin, 40, 41
Thorax, **105**
Threshold/threshold stimulus
  muscle contraction and, **18**
  neuron action potential and, 39–40
Threshold voltage, **40**
  for muscle contraction, **18**
  for neuron action potential, **40**
Thymus, 58*f*
Thyroidectomy, 59
Thyroid gland, 58, 58*f*
  metabolism and, **58**–62
Thyroid hormone (thyroxine/T_4), 58, **58**
  metabolism/metabolic rate and, **58**–62
Thyroid-stimulating hormone
    (TSH/thyrotropin), **58**
  metabolic rate and, 60–61
Thyrotropin-releasing hormone (TRH), **59**
Thyroxine (T_4/thyroid hormone), 58, **58**
  metabolism/metabolic rate and, **58**–62
Tidal volume (TV), **104**, **105**
TLC. *See* Total lung capacity
Tonicity, **7**
Total force, muscle contraction, **25**
Total lung capacity, **105**
Transducers, force, **15**
Transduction, sensory, **37**
Transfusion reactions, 165
Transmembrane proteins, in glucose
    reabsorption, 138
Transport, cell, 1–14
  active processes in, **1**, 2
  passive processes in, **1**
Treppe, **20**
TRH. *See* Thyrotropin-releasing hormone
Trigger zone, **39**
Triglycerides, **125**
Tropic hormones, 59, 67
T score, 65
TSH. *See* Thyroid-stimulating hormone
TTX. *See* Tetrodotoxin
Tubular reabsorption, **130**, **137**, **138**
  glucose carrier proteins and, **138**–140
  urine concentration and, 137–138
Tubular secretion, **130**
TV. *See* Tidal volume
Twitch, muscle, **16**–18, **18**
Type 1 diabetes mellitus, **62**
Type 2 diabetes mellitus, **62**
Type A blood, 165, 166*t*
  inheritance of, 165
Type A blood antigen, **160**, 165, 166*t*
  antibodies to, 166, 166*t*
Type AB blood, 165, 166*t*
  inheritance of, 165
Type B blood, 165, 166*t*
  inheritance of, 165
Type B blood antigen, **160**, 165, 166*t*
  antibodies to, 166, 166*t*
Type O blood, 165, 166*t*
  inheritance of, 165

Unfused tetanus, **22**, 23
Uniporter, 11
Urine concentration, solute gradient and, 137–138
Urine formation, hormones affecting, 140–141
Uterus, oxytocin affecting, 58

Vagal escape, **93**
Vagus nerve (cranial nerve X), heart affected
    by stimulation of, 93–94
Valves, of heart, 84

Variable region (antibody), 180
Vasoconstriction, **74**
Vasodilation, **74**
VBD. *See* Vertebral bone density
VC. *See* Vital capacity
Velocity, shortening, **26**
Ventilation (breathing), **103**, 104, 104*f*
  acid-base balance and, 148, 148–150, 152–154
  frequency of, **104**
  mechanics of, 103–115
  muscles of, **104**, **105**
Vertebral bone density, 65
Vesicular transport, **2**
Vessel length, blood flow and, 74, 77

Vessel radius
  blood flow and, 74, **74**–76
  glomerular filtration and, 130–133, 135–137
  pump activity and, 80–81
Vessel resistance, **73**, 74
  blood flow and, 73–74, 74, 74–86
    compensation/cardiovascular pathology and, 84
  blood pressure and, 73–74, 74
Viscosity, 76
  blood, 74, 76
  blood flow and, 74, 76–77
Vital capacity (VC), **105**
  forced, **105**

Voltage (stimulus), **18**
  muscle contraction affected by, **18**–20
Voltage-gated potassium channels, refractory periods and, 42
Voltage-gated sodium channels, **40**–42
  refractory periods and, 42
Volume, osmosis affecting, 6
Vomiting, metabolic alkalosis and, 152

Wave summation, **20**, **92**
Weak acid, **147**
Weak base, **147**
Western blotting, 182–184